# ARTIFICIAL
# INTELLIGENCE IN
# LIBRARIES AND
# PUBLISHING

Charleston Briefings: Trending Topics for Information Professionals is a thought-provoking series of brief books concerning innovation in the sphere of libraries, publishing, and technology in scholarly communication. The briefings, growing out of the vital conversations characteristic of the Charleston Conference and Against the Grain, will offer valuable insights into the trends shaping our professional lives and the institutions in which we work.

The Charleston Briefings are written by authorities who provide an effective, readable overview of their topics—not an academic monograph. The intended audience is busy nonspecialist readers who want to be informed concerning important issues in our industry in an accessible and timely manner.

Matthew Ismail, Editor-in-Chief

# ARTIFICIAL INTELLIGENCE IN LIBRARIES AND PUBLISHING

WITH CHAPTERS BY:
DANIEL W. HOOK AND SIMON J. PORTER
CATHERINE NICOLE COLEMAN AND
MICHAEL A. KELLER
JAMES W. WEIS AND AMY BRAND
RUGGERO GRAMATICA
HARIS DINDO
TODD A. CARPENTER

EDITED BY
RUTH PICKERING AND
MATTHEW ISMAIL

Published in the United States of America by
ATG LLC (Media)
Manufactured in the United States of America

DOI: https://doi.org/10.3998/mpub.12669942

ISBN 978-1-941269-54-1 (paper)
ISBN 978-1-941269-55-8 (ebook)
ISBN 978-1-941269-56-5 (open access)

http://against-the-grain.com

# CONTENTS

# INTRODUCTION
## Ruth Pickering

When I was first asked to write this Charleston Briefing, the words of Bill Gates sprang to my mind, "In 5 years from now, we will all laugh about what we have today." Technology is such a fast-moving space that writing about it carries the risk of irrelevance in a very short period of time.

When we started writing this Briefing, no one had heard of Covid-19, but before it has even been published, several vaccines have been developed.

What times we are living in.

We are all overwhelmed by the amount of information that exists. According to a recent Gallup poll, 58 percent of Americans said that the increase in news sources makes it harder to stay informed. We are all readers and users of content, and some of us also own and curate content collections. Whichever space you occupy, the challenge of the ever-increasing fragmentation of content across different domains, formats, and languages is not new. The aim of this briefing is to provide some ideas on how artificial intelligence (AI) can help make content more discoverable, and to enable the extraction of insights and non-obvious connections at scale.

No one person or organization can answer this question, so we decided to collaborate with a group of experts from different areas, and I would like to thank all the chapter authors for their willingness to participate and for their ideas and perspectives.

Please join me several years from now in Charleston, at 4 P.M. on a Tuesday in early November, to share a glass of wine and discuss what AI might look like in the future for libraries and publishers. Let's compare what it looked like while standing in 2021 when AI was still in its infancy but was already doing great things.

# CHAPTER 1

# AI VERSUS IA

Daniel W. Hook (0000-0001-9746-1193) and
Simon J. Porter (0000-0002-6151-8423)

"You just can't differentiate between a robot and the very best of humans."

—Isaac Asimov, *I, Robot*

## INTRODUCTION

Let us imagine a world in which artificial intelligence (AI) is deployed ubiquitously across the research ecosystem: As you perform your research tasks, such as carrying out a literature search, compiling ideas, taking notes, summarizing and classifying relevant material, and so on, an ever-present set of AIs help you work faster, more accurately and more efficiently. If your research has a practical element, an AI helps take readings, suggests likely parameter spaces to explore, or analyzes and classifies your data outputs. When you write up your research, a range of tools might be available to you to improve the quality of your communication, assess your work, and give you feedback so that you can improve your manuscript before submitting for peer review. As a peer reviewer, an AI can help contextualize the piece that you're reviewing, just in case it is not at the center of your work. Perhaps it is one of a raft of new interdisciplinary pieces that are crossing your desk and which are challenging to review because, while you may be an expert in one of the research areas of the paper, you need to understand the interrelation of the work with less familiar areas in order to formulate an opinion.

Most researchers would recognize the value of such tools, but only a small subset of researchers would recognize that all the tools mentioned before are examples of AI-driven tools that already exist. In a future-looking article such as this one, we are bound to ask: What will be the capabilities of the AIs of the future? Will they be able to perform whole experiments? Many of the tasks that we have mentioned before are tasks that postdocs and PhD students currently perform for many academics. Will there be less need for such roles? What will that do to an academic education and the path to research? Will we need researchers at all? To address these big questions in this article, we will look at some of the bigger picture questions around AI, not just in a research context but in a broader societal context, which we can then relate back to research.

This chapter is arranged into eight subsections. First, in "What Is Artificial Intelligence?" we make a high-level definition of the topic as background context for the following sections. We then turn to a historical perspective in AI versus history before examining how the revolution in AI could change our economy in AI versus economics. All the different streams in this chapter are intended to relate to how AI can affect and interact with the research world, and we introduce those issues more cogently in AI versus research and continue to think about the critical relationships between humanity and AI in AI versus trust. AI is becoming ubiquitous around the world and will have impacts throughout society—we examine the importance of AI as a productivity tool in AI versus the workplace and finally consider whether AI can really master that most human of acts, innovation, in AI versus creativity before making some concluding remarks.

## WHAT IS ARTIFICIAL INTELLIGENCE?

The term "artificial intelligence" is among those contemporary buzzwords that have moved from the tech industry out into the everyday world. In the start-up sector, making sure that your shiny new start-up company features AI on your website (ideally, alongside "blockchain," "adaptive learning," and "personalization") or better yet has a. ai domain name, will instantly increase the interest (and potentially the valuation) of your enterprise.

This is, of course, just one part of AI in our consciousness, as we are often exposed to the idea both in fiction and in news, with headlines that suggest that AI is either here to destroy or replace us. From the deeply troubled and conflicted AI HAL in Clarke and Kubrick's *2001: A Space Odyssey*, to the *Terminator*'s humanity-destroying Skynet, and recent predictions by Elon Musk that AI will eventually overrun humanity (Dowd 2017), popular images of AI are often negative, tending to be extreme or dark caricatures. However, they are neither easy to, nor should they, be dismissed out of hand. The potential of AI to change our economy, our society, our relationships, and our perception of the world is very real—whether it will be for better or worse is unclear because, in the same way mobile devices have changed social interaction, the way in which we let AI augment our lives may not necessarily be universally positive.

Humans are both creators and users of tools. This fundamental characteristic has led to our dominance in the Earth's ecosystem. In light of the large-scale problems that we face today as a society, whether it be global warming, threats to liberal democracy, poverty, or any one of many others, AI can potentially be of assistance. Yet AI may itself be a challenge to us—not because an all-powerful AI will seek to subjugate humanity, but because we may become at least as dependent on AI as we are on our mobile devices. Will AI technology be ubiquitous yet unseen, supporting and enhancing our world, or will it lead to greater inequality between rich and poor, drive divisions in society, and cause us to question the reality around us?

In this piece, we wish to draw a distinction between artificial intelligence and intelligent augmentation. We argue that the current state of development of AI as a field is very much aligned with the latter and that the former is still centuries away. Seen through this lens, we can stop worrying about Skynet and focus on how we want these technologies to shape our society, our workplace, and our relationships.

In this chapter, we will not focus on what is variously called strong AI, or generalized AI. Rather we will explore the subtler forms of AI that already exist in the world around us today and those which inhabit our near futures.

## AI VERSUS HISTORY

It is often said that "it takes a long time to be an overnight success," and that certainly holds true for AI. Originally developed by researchers in the late 1950s, the ideas behind machine learning were pioneered by such greats as Arthur Samuel, who coined the term *machine learning*; Norbert Wiener, widely regarded as the father of cybernetics; Marvin Minsky; Allen Newell; Herbert Simon; and others (Brockman 2019; Wiener 1954).

Spurred on by Russia's successes in its space program and, in particular, by the successful orbit of Sputnik, the United States began to spend vast amounts on research and development. The United States feared that the Russia would gain supremacy in the skies and considered this an unacceptable position. As a result, defense and non-defense R&D spending in the United States peaked at 6.4 percent of GDP in 1968 (White House 2020), pouring money into activities that were historically used to half as much funding, and where spending levels have now renormalized. This investment in the knowledge infrastructure, the areas that supported the National Aeronautics and Space Administration (NASA) and the space race, would lead not only to the Moon landings, but would also have knock-on effects across science and technology. This most notably sparked the computer revolution in the 1980s and led to the Information Age and the emergence of the Internet from its precursor at the Defense Advanced Research Projects Agency (DARPA) in the 1990s.

While machine learning was in its early stages of development during the second half of the 20th Century, there was neither yet enough computer power nor enough data for it to develop much beyond theoretical science. Enter the age of the Internet. With the vast availability of data, coupled with computer processing power at scale, AI started to become the hottest technology field and consequently one of the fastest growing research fields (see Figure 1.1).

The research "space race" between the United States and China began in 2016 as AlphaGo, a specialized deep learning AI from Google, beat the world's best Go player (Silver et al. 2016) and caused China to experience its own "Sputnik moment" (Lee 2018). It might be surprising for a game to spark such an extreme response; however, the game of Go occupies a special place in Chinese culture. It is viewed not only as a game but as an art form, and one that all intellectuals in China would be expected to master.

The technology that led to this pivotal moment for China had its roots in an earlier human–computer conflict less than 20 years before: namely the Kasparov–Deep Blue chess match. In comparison with AlphaGo's

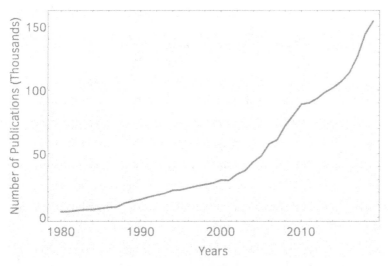

**Figure 1.1. Number of publications (articles and conference proceedings) published between 1980 and 2019 that are classified in ANZSRC Field of Research Code 0801: Artificial Intelligence and Image Processing.**
*Source:* Dimensions.

performance, the scale of Kasparov's defeat at the hands of IBM's Deep Blue chess computer is marked. Kasparov was unarguably one of the strongest chess players ever to have lived and took on Deep Blue in New York in 1997—the game was a rematch from an encounter one year earlier, when Kasparov had beaten the Deep Blue in Philadelphia four matches to two. The result of the 1997 rematch was widely reported as a significant step for machine learning and AI, and indeed it was. But the success was on a different order to that achieved by AlphaGo almost 20 years later. Deep Blue beat Kasparov by just one match in a set of six: 3½–2½. In the year between Kasparov's win and Deep Blue's slim victory, the team at IBM had not specifically improved the computer's ability to play chess, but had focused their efforts on teaching the computer how to beat Kasparov himself—an approach that was highly controversial at the time (Kasparov and Greengard 2018). Kasparov was not merely attempting to beat a generalized chess computer but one that was specifically optimized to beat his particular playing style. When, in March 2016, AlphaGo beat Go world champion Lee Sedol, it won by four games to one, having taught itself to play Go with absolutely no built-in optimization for Sedol's playing style.

The advance in AI and particularly in convolutional neural networks is impressive. However, they still require a vast amount of data and computer power at scale to perform these feats. We have AIs that are getting extremely good at specific tasks. In some tasks AIs are already better than humans, and in a few more years they will be significantly better than humans in a much wider range of activities. We are still a very long way from generalized AI— intelligence that can reason across multiple distinct, different, and unrelated tasks. Yet it is clear that, even at its present stage of development, AI can already be thought of as an intelligence-enhancing tool, not only in simple tasks but also in complex and subtle tasks. In very narrow fields, AIs are now better pattern matchers than the most expert humans.

## AI VERSUS ECONOMICS

There is some contention on the numbering, but the rise of AI and its associated technologies, such as robotics, is often described as either the Third or Fourth Industrial Revolution. The First Industrial Revolution can be uncontroversially thought of as the period after 1750 in which steam power was harnessed. The Second Industrial Revolution is often thought of as the period from 1870 up until the beginning of World War I and concerned the widespread utilization of electricity and mass production. Some date the beginning of the Third Industrial Revolution in the period immediately after World War II, with the advent of the nuclear age and the rise of general-purpose electronic computers. This would then make the Fourth Industrial Revolution, also referred to as "Industry 4.0," as the current period comprising the rise of the Internet. Others refer to this era as the Third Industrial Revolution and don't credit the post-war period, as the science did not translate to a significant technological shift in industrial processes.

Each industrial revolution to date has led to significant upheaval in the workplace for the majority of the population. As technology replaced activities in the workplace that either required low-level physical or mental capability, the work that remaining workers carried out was of a high level, often requiring new skills. Those whose jobs were replaced needed also to learn new skills either in the same or different professions (Avent 2014). In each revolution, whole new classes of jobs have also sprung up as a result of technological

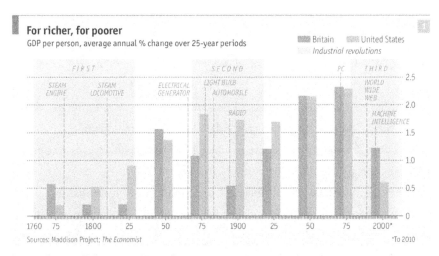

Figure 1.2. Time delay for society to experience the positive financial effects of an industrial revolution (Avent 2014).

developments. Yet, these changes all take time and there are social and economic frictions along the road. Figure 1.2 shows explicitly the time delay between the first innovations in each revolution and the time for the GDP of either Britain or the United States to respond.

If we look positively on the effects of AI, it is likely that the current industrial revolution will have the same effects of increasing the wealth and prosperity of our society. There is, however, a more extreme dynamic at play in the AI revolution. Whereas previous revolutions have provided a step up in the tools that they have provided and made workers directly more effective, this the first revolution where the tools have the ability to improve themselves, building better tools. A further aspect of this AI-powered industrial revolution is economic in nature: tools and the efficiencies that they bring are no longer limited to the tangible world but also act in the intangible economy (Haskel and Westlake 2018). This confluence of effects makes the industrial revolution that we are currently experiencing feel like an "exponential" revolution with technology often feeling as though it is outpacing the populace that is using it. Part of this perception may be driven by multiple areas in which innovation is happening and the fact that innovation is not limited to the industrial sector but is simultaneously taking place in the financial sector.

Indeed, in the current revolution, not only are we participating in the revolution as the inventors but also as both the input and product of the revolution. Much of the gig economy repositions workers not principally as service providers or creators, but rather as producers of the raw material that powers the new revolution—data. Likewise, consumers are no longer at the top of the pyramid as they also create data for the AIs to consume (Crawford et al. 2019; Zuboff 2019). There is a social toll to be paid in this revolution that previous revolutions have not exacted: privacy.

Previous industrial revolutions have eventually led to increased prosperity across society. However, the multifaceted nature of this current revolution gives it a different quality to those that came before. This revolution touches personal freedoms, repositioning us not only as inventors but also as the raw material and the product. The technology that powers this revolution is not only changing what happens in our workplace but also redefines the workplace itself. The pervasiveness of technology allows our workplace to expand spatially, temporally, and mentally.

It is difficult to argue that the current industrial revolution, as defined by the rise of the intangible economy, the pervasive use of technology, and the redefinition of the workplace, is not responsible for growing wealth inequality, especially in the developed nations. Certainly, the effects of technology on the economy will outpace the effects of globalization, which may now be in decline. The conflation of these effects may well cause political turmoil for an extended period as society at large fails to pinpoint the drivers of change. AI has already increased the rate of that change, and as the technology develops further, we believe that the effects may become more pronounced. We are already in a world in which data inequality leads to wealth inequality, and since the success of AI is fundamentally linked to the variety and quality of data that it has to consume, the equivalence of data and wealth in economic terms is almost certainly assured both on an individual and national level (Lee 2018).

As a final thought in this section, it is interesting to note that the economist Keynes perceived technology as a route by which humanity would be freed from work and that the problems that we would face at this stage in our development would be centered around finding meaning in our lives since all the basic requirements of survival would already be met through the economic development of our society (Keynes 1932; Bregman and Manton

2017; Susskind 2020). Keynes could not have foreseen the level at which quality of life has improved globally in the intervening 90 years (Rosling, Rosling, and Rosling Rönnlund 2018), the expansion of the population at the level which has come to pass, and the effects of technology on our existence. Nor could he have anticipated the effects of the most recent revolution on wealth inequality both between countries and individually. Although we have focused on the AI revolution in this piece, a counter-balancing revolution may already be underway in the shape of a new approach to economics codified in the sustainability revolution (Raworth 2017).

## AI VERSUS RESEARCH

Research into AI is continuing apace, as seen in Figure 1.1. Over one million papers have been written in the field in just the past decade. With the emergence of deep learning and the translation of the technology from the lab to a ubiquitous worldwide network of connected devices from which data can be used to further fuel both usage and research, there is no sign that this rise will slow down any time soon. To understand the landscape better, Digital Science employed its own machine learning algorithms to topic map and classified the research that has been carried out in the field over these one million recent papers.

Dimensions, Digital Science's scholarly information database, contains around 107 million scholarly articles, chapters, conference proceedings, books, and preprints at the time of writing this article. These articles are automatically classified, using machine learning, against several different classification schemas (Hook, Porter, and Herzog 2018). For the purposes of the analysis presented here, we used the Australian and New Zealand Standard Research Classification (ANZSRC) for fields of research (Wikipedia 2019) and took articles that fell into the "0801 Artificial Intelligence and Image Processing" classification as the subset of publications for our analysis. We then performed a more detailed clustering of these results into topic groups. The topic map process made use of abstract texts but not full texts for this analysis. We looked for a natural emergence of 10 to 20 subgroups that had approximately similar distinctness in nature, and then named the groups based on the most prevalent noun phrases that had been extracted as part of the topic mapping process. The 15 resultant groups are shown in Table 1.1.

**Table 1.1. Sub-classification of ANZSRC code 0801 into 15 topics, de-rived from a topic mapping approach on Dimensions data holdings.**

| Number | Primary Topic Group |
| --- | --- |
| 0 | Signal Processing |
| 1 | Control and Filtering Systems |
| 2 | Power and Energy Systems |
| 3 | Video, Image, and Wireless Systems |
| 4 | Scheduling, Networks, Agent-based Systems, and User Behaviour Prediction |
| 5 | Hardware and Software Implementations, Real-Time Systems |
| 6 | Data Mining, Training, Prediction, and Forecasting |
| 7 | Fuzzy Logic and Clustering |
| 8 | Semantic Analysis, Search, and Retrieval Systems |
| 9 | Image Segmentation and Colour Processing |
| 10 | Computer Vision and Object Tracking |
| 11 | Vehicles, Traffic, Sensors, and Sensor Fusion |
| 12 | Classification, Identification, and Recognition Systems |
| 13 | Robot navigation, Movement and Control, Human–robot Interactions |
| 14 | 2-D/3-D Imaging Systems, Image Reconstruction |

To understand the international characteristics of AI research, we then mapped each paper in the clustered set into one or more of the 15 categories in Table 1.1 while also mapping each paper to the countries associated with the institutions of the co-authors of the paper through the Global Research Identifier Database (GRID). The result of the analysis is shown in Figure 1.3.

It is interesting to note the different focuses of different countries in the research that they are funding. From Figure 1.3 it is very easy to see that category 4, "Scheduling, Networks, Agent-Based Systems and User Behaviour Prediction," is an area of intense work with only Japan not favoring this area. Japan, however, has specific strength in category 13, "Robot Navigation, Movement, and Control, Human–robot Interactions," which makes sense given their long-time investments in robotic research. Russia, on the other hand, has specific strength, relative to its other research intensiveness, in category 5, "Hardware and Software Implementations, Real-time Systems." What is less instantly evident from Figure 1.3, however, is the significant strength of China in almost every field. If we were to replot Figure 1.3 as a ranking based on the volume of output, then China would hold first place in 12 out of 15

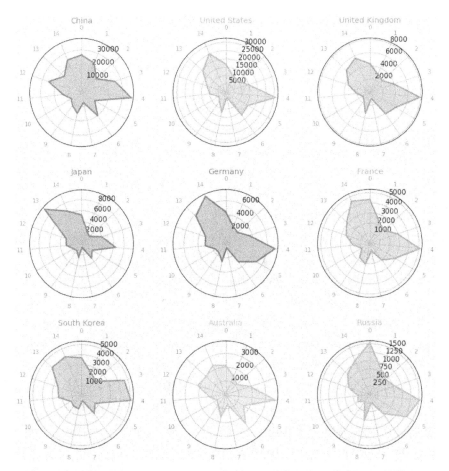

**Figure 1.3. Mapping of AI research into the classification shown in Table 1.1. The nine countries shown here are selected by the size of research output and likely interest to the reader. The scale on each radar plot shows the number of publications produced between 2010 and 2019 in each of the 15 classifications in Table 1.1. Note that each scale is different.**

*Source:* Dimensions.

classifications and second place in the remaining three classifications. Indeed, if we look at the most connected institutions in the world of AI research, it is clear that Chinese research institutions are not simply among the best in the world but currently have the majority of the top institutions.

## AI VERSUS TRUST

No discussion about AI is complete without at least a brief examination of the many concerns that lie at the center of this technology. By employing the concepts of deep learning, we are moving toward self-learning systems, the function of which we don't fully understand. In the same way that most students are no longer taught how to do long division by hand, instead relying on their pocket calculator, computer, or mobile phone, we are beginning to rely on black box technologies; we cannot easily see inside the "mind" of the computer, so we need either to find a route to unpick the thinking of the computer, or we need to be able to trust it. In essence, we are looking for a modern-day equivalent of Asimov's famous three laws of robotics (*Handbook of Robotics, 2058 A.D.* in Asimov 1942).

The development of AI in our society has been mainly innocuous—it has been a quiet (perhaps even a silent) revolution. Most people do not think of Uber as an AI company, but the routing of cars and drivers is a complex problem of optimization that benefits from learning algorithms and data. Most people don't think of Google Translate as an AI system, yet, quietly on September 27, 2016 (Google's 18th birthday), the company replaced the rules engine on which the system had previously operated with a machine learning algorithm that improved its translation capability based on the exposure to users doing translations in the system (Metz 2016). Google Health has now developed the capability for an AI to detect cancer from a scan more accurately than any human (Griffin 2020). These technologies seem wondrous, but one should not assume that they are unbiased.

If one does not understand the technology, then it is difficult to believe that a simple learning system can be biased. After all, the algorithm may be transparent and easy to understand, often almost trivial. The system then learns and develops itself. The critical component is the data that are fed into the system. By 2020, the fallacy of bias in AI is now well known with a number of high-profile examples, including a "racist" soap dispenser (Dennis 2018) and an automated recruiting tool that hired only men (Reuters 2018), to mention just two examples that have catapulted AI ethics into the limelight (Levin 2019).

Issues around AI are becoming a research hot topic with institutes not only popping up around the development of technology but also those that do research on the social and ethical implications of AI such as New York University's AI Now Institute (Whittaker et al. 2019). As we write this chapter, there are many live discussions concerning the regulation of AI—in 2019, the European Commission published the guidelines for "Trustworthy AI" and in the United States, the "Algorithmic Accountability Act" is under discussion to ensure that bias and equality are at the center of legislation.

It is issues such as these, hidden deep inside the data that feeds an AI, that make for disturbing reading. Unerringly prophetic in his books 2001 and 2010, Arthur C. Clarke already saw that an AI given conflicting instructions or being fed poor data would lead to unforeseen consequences. From a technical perspective, researchers such as Been Kim at Google Brain are working to find ways to look inside the black box so that we can begin to understand the biases in the data (Pavlus 2019). Hopefully, within a few years, "Testing with Concept Activation Vectors" (TCAV) or similar technologies that are being developed will allow us to have confidence in AI systems.

Humans have been the ultimate pattern-matching machines on the planet, but AIs are now taking that crown. On the one hand, this is a key part of what makes AIs appealing—they are tools that facilitate enhanced perception beyond standard (or even honed) human capability; the ability to parse data faster and more completely; or, the ability to perceive patterns in vast seas of information. And yet, these tools are far more subtle than those of the past, and can fail us in much more subtle and difficult-to-detect ways. In terms of trust, humans, as imperfect as they are, operate with reference to an ethical framework. A parallel needs to be developed for AIs. Of course, humans can and indeed do choose to move outside ethical frameworks and hence trust, and ethics may ultimately be an area in which AIs truly and necessarily exceed humans in capability.

## AI VERSUS THE WORKPLACE

More locally, what effects can we imagine in the workplace? A recent report by PwC (2018) gives an interesting insight into the future of the workplace. The report attempts to quantify the problem of time delay mentioned earlier by classifying the effects of AI in the workplace into three waves:

- Wave 1: Algorithmic (early 2020s): Automation of simple computation tasks and analysis of structured data affecting data-driven sectors such as financial services.

- Wave 2: Augmentation (late 2020s): Dynamic interaction with technology for clerical support and decision-making. Also includes robotic tasks in semi-controlled environments such as moving objects in warehouses.
- Wave 3: Autonomous (mid-2030s): Automation of physical labor and manual dexterity, and problem-solving in dynamic real-world situations that require responsive actions, such as in transport and construction.

These different waves hit different industries to different extents and, consequently, different countries and different job types will experience disruption at different rates and at different times (see Figure 1.4). Clearly the first wave, which we are experiencing now, only has a comparatively minor impact on our everyday working lives, but the second and third waves are much more pronounced in certain fields. The first wave has already brought us a number of innovations, and consequently disrupted a number of jobs. The AI revolution has been a silent one in many respects: AIs now land planes, optimize delivery routes, optimize pickups for online taxi services, and check for plagiarism in student work. While all these examples (possibly aside from landing a plane) seem quite innocuous and probably the types of things that

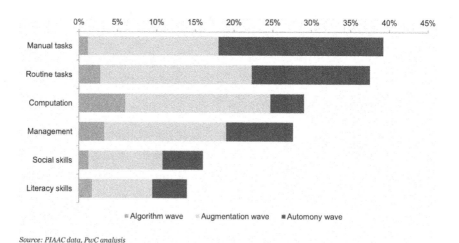

*Source: PIAAC data, PwC analysis*

**Figure 1.4. Proportion of tasks from jobs at high risk of replacement by AI-related technology. Reproduced from (PwC 2018).**

we all imagined AIs to be good at a few years ago, there are some more startling technologies that are now emerging.

Of particular note are online chat systems that can now answer your customer support queries having learned from human support agents (Walch 2019). This type of interaction has been imagined since the time of Turing, when he developed his famous test (Turing 1950). AIs are now smart enough to interact with us, without us being able to tell the difference between an AI and a human in simple conversations around specific topics. As we move from an era in which AIs optimized the routing for car services like Uber and Lyft into an era in which the AI is driving the car, the ubiquity of AI will transition from a hidden low-key state to a very evident replacement of humans by machines and algorithms. This is when we should expect the tension between intelligent augmentation (rather than a nebulous concern about AI) to become commonplace. This will parallel the stress on society induced by previous industrial revolutions.

Paradoxically, as the world of research is often the early-stage innovator with many technologies, they are also often slow adopters of new technology in their work practices. Researchers who write algorithms and AIs to search datasets to identify patterns tend to feel safe in their jobs, confident that the work that they do is not duplicable (Figure 1.5; Munroe 2017). Indeed, many of the types of tasks shown in Figure 1.4 are not immediately relevant in most research contexts. However, elsewhere in academia, there are significant opportunities for AI, and potentially a leaf can be taken from the business world.

**Figure 1.5.  xkcd commentary on the use of AI in solving research problems (Munroe 2017).**

The work of Davenport and Ronanki (2018) suggests three types of AI—most of which fall into the first two classifications from the PwC study. These are (i) process automation, (ii) cognitive insight, and (iii) cognitive engagement. In a business context, process automation centers around things such as billing and checking data transfers, however in an academic context, the focus would be on data cleansing or processing—precisely where it is currently used most. Outside the research process, in the wider university context, one could imagine this type of technology being used by the teaching parts of an institution to ensure that student details were kept up-to-date, or for checking that progression, attendance, and engagement were optimal. In a library domain, some of the most interesting use cases rest in cognitive insight classification and might include optimization for content acquisition, content discovery, or content delivery. In the cognitive engagement class, intelligent search support, helping students to understanding referencing styles, suggesting references (e.g., hamlet at MIT)[1], and many other use cases spring to mind.

In the research context, there are now a large number of tools with a vibrant innovation ecosystem growing up around AI tools that support research (Webster 2019). These are not only located in long-anticipated areas around research discovery, such as Yewno,[2] Iris.ai,[3] Wizdom.ai,[4] Diffeo,[5] Meta,[6] and Dimensions, but also around the following:

- research classification systems (as shown in the previous section with Dimensions);
- AI-enhanced peer review, such as the new Artificial Intelligence Review Assistant (AIRA) system introduced by Frontiers in 2018 (Frontiers Announcements 2018);
- systems that help researchers write more clearly and naturally by applying AI to learn from published scholarly papers so that the advice given is specific to research writing such as Writefull;[7]

---

1  https://hamlet.andromedayelton.com/
2  https://www.yewno.com/
3  https://iris.ai/
4  https://www.wizdom.ai/
5  https://www.diffeo.com/
6  https://www.meta.org/
7  https://writefullapp.com/

- systems using researchers and experts to train AIs, which are then able to automatically tag data. Research projects such as the citizen science galaxy classification project, Galaxy Zoo, have developed impressive technology in one area that have then been generalized and commercialized in products such as 1715Labs[8] (Fortson et al. 2012);
- a robotic lab with a cloud-based user interface that uses AI to search the parameter space of outcomes to find optimal results, such as Transcriptic (now Strateos)[9] (Bates et al. 2017);
- using machine learning approaches to understand what makes an article reproducible and then allows a researcher to understand if the explanation of their research provided in their paper is likely to be reproducible, such as Ripeta;[10]
- automated ideation systems that help humans to generate innovative ideas, such as Iprova;[11] or
- leveraging mobile devices and smart speakers to allow a virtual lab assistant to support your work, such as in the work of Sartorius (Sartorius 2019).

All of these areas will improve different aspects of the reproducibility, robustness, and communication of research, but this is only the technology of the first wave of automation—the algorithmic approach. Within the next 10 years we are likely to see smart lab assistants that are much more multifaceted than those currently available, as well as systems that are able to assess the novelty, correctness, or potential for the translation of a piece of research by some measure.

There are many concerns with such systems becoming commonplace in specific areas of the research ecosystem. While these tools can be extremely helpful in extending human cognition in the lab, we need to be careful that they do not become inappropriately used to replace the roles played by early career researchers. There is an important balance to be struck for these newer researchers between learning and interfacing efficiently with new technologies

---

8  https://www.1715labs.com/
9  https://www.transcriptic.com/
10  https://www.ripeta.com/
11  https://www.iprova.com/

and being able to understand the key skills that underlie the use of such technologies. In some disciplines, it may be highly appealing to use AI-based systems instead of postdocs for some activities. If this logic is applied too readily, perhaps we will fail to value and hence train the next generation of researchers, narrowing the funnel of young researchers too quickly and ending up with too few researchers in particular fields. At the same time, we already see the challenges in academia in keeping researchers with specific skills such as data science. There is significant potential for high pay outside academia for individuals who understand data science. Such skills have not been highly valued in academia and hence a generation has been lost to industry.

Research evaluation is another area that must be treated with care. While AIs may be good at providing suggestions for reviewers of grants and papers, and may in the near future be able to assess the novelty and impact of a piece of research, like the drive for metrics in the United Kingdom 10 years ago (Wilsdon 2016), the excessive use of AI could threaten to narrow research and stem diversity—rating highly only the research that conforms to certain learned templates. Of course, such an outcome would be dependent on the data that would have been fed into the AI. Detecting bias in this context would however potentially be an order of magnitude harder than detecting bias in the ethical cases discussed before. Research tends to be a long-range field in which developments can take years to prove. As a result, the feedback loop can be extremely slow, and this seems to be a weakness for the creation of a research evaluation AI.

To summarize this section, Table 1.2 gives examples of some of the AIs mentioned in this section according to the PwC framework.

## AI VERSUS CREATIVITY

The final stop on this whirlwind tour of AI is to consider intelligent augmentation in the context of creativity. Creativity in humans is, for most, a private experience and is often considered to be rare. Few have documented their creative experience in detail but Fields Medalist Cédric Villani gives a very personal account (Villani 2015), and Joao Magueijo writes about how he believes a love-hate relationship is necessary for the most innovative of ideas (Magueijo 2004). It is difficult to believe that this most human of experiences

**Table 1.2. Examples of AI systems and the different roles that they can play in scholarly research. Grey squares are areas that have not yet been significantly covered; pink squares are those where we potentially should not use AI.**

| Role | Algorithm | Augmentation | Automation |
|---|---|---|---|
| Practice of Research | Checklist-type systems | AI Lab Assistant | Transcriptic |
| Analysis | R, Mathematica, Python | 1715Labs | |
| Authoring | Paper Digest | Writefull | Beta Writer |
| Peer Review | Dimensions | Dimensions | |
| Discovery | Dimensions, Yewno, Iris | Dimensions, Yewno (Automated Classification) | |
| Evaluation/Policy | Impact Factor | Dimensions | |
| Translation | | Iprova | |

can be reduced to pattern matching. Can the intellectual leaps that appear to be required for the most fundamental research problems be codified in a computer?

In other areas of expertise there are significant developments in achieving some form of creativity with computers. For example, AIs can now create jokes, some of which are truly funny (Barber 2019). It appears that the technique in this case relies on the understanding of word plays that might be perceived as funny—existing material and pattern matching is sufficient to create lines that are moderately amusing. In 2020, Beethoven's 10th Symphony will be performed as newly completed by an AI (Delbert 2019). Yet, on closer reading, it is clear that the AI really is acting as an IA, helping human composers with suggestions and contextualizing from Beethoven's work—pattern matching and extending human cognition on a grand scale, yet with the creativity remaining human.

In 2019, the academic publisher Springer Nature launched the first machine-written academic text—a review of literature on lithium-ion batteries (Writer 2019). The process of writing was, again, based on pattern matching and categorization technology, with the book itself having a foreword that explains the process of writing. This is already an impressive feat, especially since the comments from the external (human) peer reviewers were positive, commenting only that the text was occasionally a little dry. However, it continues to be a work that is fundamentally not novel in its creativity.

## CONCLUDING DISCUSSION

Periods of industrial revolutions often define turbulent times, and as fast as technology may appear to be moving forward now, we are only at the very beginning of the AI revolution. The impacts of today's opportunities and pitfalls both sociological and technological are as yet unclear. As Nobel Prize winner, Niels Bohr famously said, "Prediction is very difficult, especially if it's about the future," and despite the dire pronouncements of Elon Musk, humanity is capable of making intelligent decisions to try to nudge us toward a future in which we continue to see improvements in the living standards and well-being for all.

The music industry has undergone several revolutions of its own over the past century, first with the introduction of sound recording and subsequently with the advent of live streaming. Any future gazer of a 100 years ago might well have predicted the end of the choir, the end of the orchestra, and the end of music stars, and yet, while the economics of the music industry are very different to those a 100 years ago and different again from how they were just 10 years ago, we still have singers, instrumentalists, and groups who thrive (Krueger 2019). There is a human factor in the technologies that we choose to adopt.

The current level of development of AI is impressive and the tools that are being produced certainly qualify for the name of "intelligent augmentation." The possibility to extend the range of human cognition and perception with the tools that are being developed is awesome and yet, at the same time, there are clear limits and dangers to this technology.

In a research context, smart individuals adopt new technologies with care and healthy skepticism. There are areas where AI will help, others where they will hinder, and yet others where our current insights leave us without a clear determination of whether intelligent augmentation, let alone a strong AI, can ever be developed. In the current revolution, it appears that while these tools will change many aspects of human society and extend our capabilities into areas we cannot yet guess, it is likely that neither IA nor AI will duplicate or exceed our ability to think beyond the patterns that we perceive—to be truly imaginative and uniquely creative.

## ACKNOWLEDGMENT

The authors would like to thank Jared Watts for his support and insight in the analysis required for the 15 topic classifications of AI in the section "Research versus AI."

## CONFLICT OF INTERESTS

Both authors of this article are employees of Digital Science. Digital Science is an investor in or owner of several products named in this article including Dimensions, GRID, Ripeta, Transcriptic, and Writefull. Digital Science is a sister company of Springer Nature, which published the first machine-written academic book.

## REFERENCES

Asimov, I. 1942. Runaround in Astounding Science Fiction. New York: Street and Smith Publications, Inc.

Avent, R. 2014. "The Third Great Wave." *The Economist* [WWW Document]. URL: https://www.economist.com/news/special-report/21621156-first-two-industrial-revolutions-inflicted-plenty-pain-ultimately-benefited (accessed January 1, 2020).

Barber, G. 2019. "The Comedian Is in the Machine. AI Is Now Learning Puns." *Wired*.

Bates, M., Berliner, A. J., Lachoff, J., Jaschke, P. R., and Groban, E. S. 2017. "Wet Lab Accelerator: A Web-Based Application Democratizing Laboratory Automation for Synthetic Biology." *ACS Synthetic Biology* 6: 167–171. https://doi.org/10.1021/acssynbio.6b00108

Bregman, R., and Manton, E. 2017. *Utopia for Realists*. London: Bloomsbury Publishing, an imprint of Bloomsbury Publishing Plc.

Brockman, J. 2019. *Possible Minds: 25 Ways of Looking at AI*. New York: Penguin Press.

Crawford, K., Dobbe, R., Dryer, T. et al. 2019. *AI Now 2019 Report*. New York: AI Now Institute.

Davenport, T. H., and Ronanki, R. 2018. "Artificial Intelligence for the Real World." *Harvard Business Review* January-February 2018: 108-116. URL: https://hbr.org/2018/01/artificial-intelligence-for-the-real-world.

Delbert, C. 2019. "Beethoven Never Finished His Last Symphony. Can Robots Complete the Job?" *Popular Mechanics*. URL: https://www.popularmechanics.com/science/a30266727/beethoven-last-symphony-artificial-intelligence/

Dennis, D. J. 2018. "AI Lacks Intelligence without Different Voices." x.ai. URL: https://x.ai/blog/ai-lacks-intelligence-without-different-voices/ (accessed February 2, 2020).

Dowd, M. 2017. "Elon Musk's Billion-Dollar Crusade to Stop the A.I. Apocalypse." *Vanity Fair* [WWW Document]. URL: https://www.vanityfair.com/news/2017/03/elon-musk-billion-dollar-crusade-to-stop-ai-space-x (accessed January 2, 2020).

Fortson, L., Masters, K., Nichol, R., Borne, K., Edmondson, E., Lintott, C., Raddick, J., Schawinski, K., and Wallin, J. 2012. "Galaxy Zoo." In *Advances in Machine Learning and Data Mining for Astronomy*, edited by Michael J. Way, Jeffrey D. Scargle, Kamal M. Ali, and Ashok N. Srivastava. https://doi.org/10.1201/b11822-16

Frontiers Announcements. 2018. "AI-Enhanced Peer Review: Frontiers Launches Next Generation of Efficient, High-Quality Peer Review." *Frontiers Research News*. URL: https://blog.frontiersin.org/2018/12/14/artificial-intelligence-peer-review-assistant-aira/ (accessed February 2, 2020).

Griffin, A. 2020. "Google AI Can Spot Possible Breast Cancer Better than Trained Experts." *The Independent. Handbook of Robotics*, 56th ed., 2058 A.D. New York, NY: Doubleday.

Haskel, J., and Westlake, S. 2018. *Capitalism without Capital: The Rise of the Intangible Economy*, Reprint edition. Princeton, NJ: Princeton University Press.

Hook, D. W., Porter, S. J., and Herzog, C. 2018. "Dimensions: Building Context for Search and Evaluation." *Frontiers in Research Metrics and Analytics* 3. https://doi.org/10.3389/frma.2018.00023

Kasparov, G., and Greengard, M. 2018. *Deep Thinking: Where Machine Intelligence Ends and Human Creativity Begins*, First edition. London: John Murray.

Keynes, J. M. 1932. "Economic Possibilities for Our Grandchildren." In: *Essays in Persuasion*. New York: Harcourt Brace, pp. 358–373.

Krueger, A. 2019. *Rockonomics: What the Music Industry Can Teach Us about Economics*. London: John Murray.

Lee, K.-F. 2018. *AI Superpowers: China, Silicon Valley, and the New World Order*. New York: Houghton Mifflin Harcourt.

Levin, S. 2019. "'Bias Deep Inside the Code': The Problem with AI 'Ethics' in Silicon Valley." *The Guardian*.

Magueijo, J. 2004. *Faster than the Speed of Light: The Story of a Scientific Speculation*. New Ed edition. London: Arrow.

Metz, C. 2016. "An Infusion of AI Makes Google Translate More Powerful Than Ever." *Wired*.

Munroe, R. 2017. *Here to help*. https://xkcd.com/1831

Pavlus, J. 2019. "Been Kim Is Building a Translator for Artificial Intelligence." *Quanta Mag*.

PwC. 2018. *Will Robots Really Steal Our Jobs?*

Raworth, K. 2017. *Doughnut Economics: Seven Ways to Think Like a 21st-Century Economist*. Chelsea Green Publishing. Hartford, VT.

Reuters. 2018. "Amazon Ditched AI Recruiting Tool that Favored Men for Technical Jobs." *The Guardian*.

Rosling, H., Rosling, O., and Roslin-g Rönnlund, A. 2018. *Factfulness: Why Things Are Better Than You Think*. London: Hodder & Stoughton.

Sartorius. 2019. "Sartorius and the German Research Center for Artificial Intelligence Launch Joint Research Laboratory" [WWW Document]. Sartorius. URL: https://

www.sartorius.com/en/company/newsroom/corporate-news/399026-399026 (accessed February 20, 2020).

Silver, D., Huang, A., Maddison, C. J. et al. 2016. "Mastering the Game of Go with Deep Neural Networks and Tree Search. *Nature* 529: 484–489. https://doi.org/10.1038/nature16961

Susskind, D. 2020. *A World Without Work: Technology, Automation and How We Should Respond.* London: Allen Lane.

Turing, A. M. 1950. "I.—Computing Machinery and Intelligence." *Mind* LIX: 433–460. https://doi.org/10.1093/mind/LIX.236.433

Villani, C. 2015. *Birth of a Theorem: A Mathematical Adventure*, First edition. London: Bodley Head.

Walch, K. 2019. "AI's Increasing Role in Customer Service" [WWW Document]. *Forbes.* URL: https://www.forbes.com/sites/cognitiveworld/2019/07/02/ais-increasing-role-in-customer-service/ (accessed February 2, 2020).

Webster, K. 2019. "Artificial Intelligence: Impacts and Roles for Libraries." Carnegie Mellon University. Presentation. URL: https://doi.org/10.1184/R1/11538750.v1

White House. 2020. "Historical Table, 9.1—Total Investment Outlays for Physical Capital, Research and Development, and Education and Training: 1962–2020" [WWW Document]. White House. URL: https://www.whitehouse.gov/omb/historical-tables/ (accessed January 2, 2020).

Whittaker, M., Alper, M., Bennett, C. et al. 2019. *Disability, Bias, and AI.* New York, New York.

Wiener, N. 1954. *The Human Use of Human Beings: Cybernetics and Society.* New Ed edition. Reprint edition, 1988. New York, NY: DaCapo Press.

Wikipedia. 2019. Australian and New Zealand Standard Research Classification. Wikipedia.

Wilsdon, J. 2016. *The Metric Tide: Independent Review of the Role of Metrics in Research Assessment and Management*, First edition. Los Angeles: SAGE Publications Ltd.

Writer, B. 2019. *Lithium-Ion Batteries: A Machine-Generated Summary of Current Research*, First edition. New York, NY: Springer.

Zuboff, P. S. 2019. *The Age of Surveillance Capitalism: The Fight for a Human Future at the New Frontier of Power: Barack Obama's Books of 2019.* London: Profile Books.

# CHAPTER 2
# AI IN THE RESEARCH LIBRARY ENVIRONMENT
## Catherine Nicole Coleman and Michael A. Keller

For those who picture well-worn books and archival collections when they think of academic libraries, it may not be immediately apparent how libraries can both benefit from and shape applications of artificial intelligence (AI). Libraries are the nexus of the preservation of memory and the future of knowledge production, both of which are key to successful applications of AI. Though the popular vision of AI is almost entirely a futuristic vision, machine intelligence is fueled entirely by past human creation and collection, whether it is remote-sensing data gathered yesterday or manuscripts from centuries past. Ordering and contextualizing that information is an essential part of knowledge creation. Speech-to-text translation, image similarity search, and word embeddings, accelerated by algorithmic processing, are already changing how we discover and interact with information online. Finding the *right* information to answer a query, finding a novel resource to drive innovative research agenda, those require the kind of human expertise found in the library and we need to bring that expertise to the design of AI.

Library collections include an extraordinarily diverse array of objects. Stanford Libraries holds Abraham Ortelius's atlas of the world, rare medieval manuscripts, Allen Ginsberg's sneakers, a piano roll recorded by Claude Debussy, the 1959 edition of *The Negro Travelers' Green Book*, as well as millions of books, journals, documents, and datasets. The hidden work that makes all these materials available to research depends upon professional staff in acquisition, information management, curation, forensics, archiving, digitization, conservation, and software engineering. While the increasing prevalence of AI in information workflows enhances discovery in exciting new ways, the long-established practices of libraries are coming to the fore as essential guiding principles for the information economy.

The users of academic libraries have needs that go beyond search. While commercial search engines apply machine learning algorithms and massive computing power to make it possible to effectively query the swamp of undifferentiated data available online, this kind of access to information on the Internet ignores context and provenance. Research outcomes depend upon reliable sources and the test of reliability requires clear context and verifiable provenance. This is true not only of pre-digital resources but also for the data that we are collecting and preserving today. Maintaining confidence in materials extends to how that information is handled as it moves from analog to digital or even from already digital into new formats and configurations. The disruption that the rapid distribution, duplication, and manipulation of content over the Internet has brought to our understanding of what is trustworthy or not is making the role of the library at this moment of extreme technological change not merely helpful but critical to our ability to make decisions and produce new knowledge.

The efficient predictive machine learning techniques behind much of AI rely upon a tremendous amount of human work, particularly in the careful curation of training data as well as choices made to define which features are of interest and what training labels should be applied. Since the predictions produced will only be as good as the training data the algorithm is fed, expertise in making those choices will produce better results. The work of catalogers, archivists, and curators all contributes valuable structure, order, and discoverability to library resources. The expert tasks of description, organization, and selection provide information that can be used computationally to train algorithms. Even more importantly, the information structuring that goes on is sensitive to the origins and context within which materials are created and collected. We have all witnessed or experienced the confusion and discord of statements taken out of context. Domain-specific contextual framing is indispensable to the heuristic of determining intent and meaning.

In the fast-moving information economy, libraries take the long view. For researchers, it can be a challenge to meet the 10-year expectation for data access that granting agencies increasingly require. And yet, so much of the critical research being done today, driven by AI, relies on looking back not only decades but millennia. Climate studies, for example, find evidence of environmental change in sources as wide-ranging as coral cores from reefs that are thousands of years old, handwritten death records from the 1800s, and the typed field notes of scientists from the previous century. It will be the same 100 years from today. Libraries have this deep-time perspective in mind when anticipating the needs of future research. Making materials available, whether they are social media streams or daguerreotype prints, involves new preservation strategies and strategies for discovery, both of which can benefit from AI. We rely on technology innovation to safeguard the past. The greatest challenge of digital preservation is managing the volatility of bits, which amounts to tracking when changes are intentional or not. Maintaining data integrity, whether the resources are complex multimedia objects, a zipped collection of files, or metadata inside an XML structure, requires redundancy and monitoring. We already use algorithms in the monitoring process, and applications of machine learning will continue to make the effort more affordable and improve reliability, speed, and accuracy.

Discovery, too, is a realm where AI will dramatically change how librarians and researchers interact with library resources. Discovery has traditionally depended upon metadata. Without metadata, the rich resources of any library are essentially hidden from anyone searching the online catalog. There is formal, structured metadata that undergirds a system of information exchange both within and across institutions. There is descriptive metadata which can take the form of annotation of an object. AI is contributing to the creation of both types of metadata. Formal rules for applying subject headings and classification systems like Dewey Decimal are particularly amenable to algorithmic processing. We have plenty of labeled data—resources already tagged—that can be used to train those algorithms. And yet, if we use only resources produced in the past to train models to label resources that are produced today, we risk overwriting the very evolution of thought and meaning that we are meant to capture. This is where the power of machine reading texts, images, and other media to identify patterns within the materials themselves, rather than imposing predefined categories, open exciting new ways to think about discoverability.

Discovery, particularly for expressive works of art and literature, but equally for any materials that have not previously been cited, has always presented a challenge. The digitization of books led to full-text search which has radically changed how we read—making it easy to target terms in context—facilitating the quick review of lengthy volumes. While it is effective when exploring the contents of an individual text, keyword search does not scale well to discovery at the level of collection, genre, or subject. For example, Stanford recently acquired a collection of more than 2,000 19th-century British novels that are relatively unknown to scholars. Since they have not been read, reviewed, or written about, metadata for meaningful categorization is sparse. Though the novels are digitized and accessible online, they are difficult to explore as a collection. A team within the library, including the curator and metadata librarian, working together with machine learning engineers, used topic analysis to "machine read" the entire corpus. The results did not align with the topic and genre categorizations that catalogers would typically use for the collection, but surfaced patterns within the corpus that point to new dimensions for discovery based not only on topics but also on style, mood, and tone. Machine learning, applied to both pre-processing and the corpus analysis, is changing the cost-benefit equation for these natural language processing techniques that were previously only feasible in a research context.

Similarity search with images is also radically changing the experience of discovery, in ways that give more control to the researcher. A photograph taken at an archaeological site can be uploaded and matched not only to other images but to an entire ecosystem of linked resources and descriptions. Existing photography collections can be filtered and sorted according to features of particular interest whether or not any metadata exists to describe that image. Metadata librarians, too, can use similarity search to de-duplicate images and uncover clues that help describe collections. We are even finding uses in conservation, where photographs of book bindings and binding repairs contain far more information for comparative analysis than could be captured in a written report. There are many different underlying techniques that drive similarity search, but the common goal is pattern matching. The pattern sought could be based on color, composition, or specific objects identified within an image. The power and flexibility of this approach is amplified when thresholds of similarity to dissimilarity across these different vectors can be applied to refine the visual query. Discovery based on patterns within content, whether visual, textual, or aural, offers new modes of access that are not dependent upon faceted search through metadata.

Another significant impact of AI will be in the process of digitization itself. Digitization has become essential for discovery, distribution, and duplication for preservation. More and more, the goal of digitization is also to make materials available for computational analysis. At one extreme are fragile treasures like the papyrus scrolls of the Herculaneum library that were carbonized by the eruption of Vesuvius. Machine learning has been applied to extract 2-D surfaces from the charred 3-D scroll so that they can be analyzed and read without having to unroll it. Libraries and archives hold troves of analog materials that are more easily amenable to digitization for computation. Analog materials from the 20th century, like reel-to-reel recordings, oral histories captured on cassette tapes, newspapers, and field notebooks are time-consuming and costly to digitize conventionally. Once digitized, generating metadata to make them discoverable adds significantly to the effort and expense. Algorithmic processing can assist in the repetitive aspects of digitization, but the greater advantage comes with the application of speech recognition to convert audio recordings to searchable text, computer vision to transcribe handwritten documents, and extracting tabular data from print documents. Converting analog collections to data opens up exciting new avenues for discovery.

At the heart of all these practical applications of AI in the library is generalized prediction. The risk in any application of predictive models on a large scale is the perpetuation of human bias in the underlying data and model. The bias inherent in machine learning is further amplified by the choices we make about how to implement AI. For example, the recursively false notion that machine-generated decisions are better, or less biased than human decisions means that, when put into practice, human discretion is given over to generalized prediction. Generalized prediction eliminates variation, privileging what is deemed most common or statistically significant. This contradicts the strengths of a research library as a holder of unique objects and a place where collections are developed to serve the changing interests of a particular research community. The challenge we face is to find the right balance by critically examining our implementation.

The quotidian work of libraries will ensure safe, trusted, and carefully curated sources for AI-driven research practices while, at the same time, opening up avenues for research through augmented discovery. Exciting new instruments to explore our cultural heritage and knowledge stores are on the horizon.

# CHAPTER 3

# ARTIFICIAL INTELLIGENCE IN THE ACADEMIC PUBLISHING ECOSYSTEM

James W. Weis and Amy Brand

## INTRODUCTION

Those of us who have chosen to devote our working lives to scholarly communication are driven by a desire to accelerate the path from research breakthrough to application and societal benefit. Yet despite the huge advances in digital publishing and the research technologies of recent years, how academics produce and consume peer-reviewed scholarship is unchanged from the print era in fundamental ways. A key reason for this is the interdependence of publishing and career advancement in academia and the ways in which the customs of the latter stifle change in the former. As a result, academic publishing practices have so far failed to take robust advantage of today's information technologies, let alone AI-based computational methods. But, little by little, publishers and aggregators are embracing new metrics, new navigation tools, and smarter approaches to content review and curation. In this chapter, we survey AI-informed developments and opportunities in each of these areas of scholarly publishing, taking care to distinguish true AI-driven approaches—systems that employ learning and other humanlike or rule-based behavior—from other computational methods.

## SMARTER METRICS

Prior to the mid-17th century, scientific communication consisted largely of personal letters between practitioners. *Philosophical Transactions of the Royal Society*, established in 1665, is widely considered to be the world's first scientific journal, underpinning a step function of ongoing refinements in scientific publishing that continues to the present. Peer and editorial review, as well as specialized journals, emerged early on as filters that allowed the growing scientific community to extract meaningful insights more easily from the increasingly broad and rapid pace of scientific research.

In this new paradigm, research impact was roughly quantified by publication output and citation-based metrics (Csiszar 2017). For individual papers, the total number of citations was (and continues to be) the most frequently used quantification of importance. For journals, Eugene Garfield proposed the "journal impact factor," a journal-level measure of per-article citation rates, which became widely adopted (Garfield 1955, 2006). For researchers, Hirsch's $h$-index attempted to "quantify the cumulative impact and relevance of an individual's scientific research output" (Hirsch 2005). As the dimensions of the scientific literature continued to grow, making personal digestion of all the literature in any field nearly impossible, administrators began to rely on these metrics for the assessment of scientific impact—which, consequently, made them targets for researchers, who need to demonstrate the scientific impact to be hired or promoted.

The transition from print-based to web-based scientific correspondences has contributed to further explosive growth in the breadth and scope of academic communication. This expansion includes not only the web replicates of traditional journal-based, peer-reviewed research articles, but also a variety of new media, including preprint servers, blogs, and even social media conversations. As a consequence, traditional citation-based impact metrics have become increasingly poor (and manipulated) proxies for actual academic impact (Wilhite and Fong 2012; Franck 1999).

While the large amount of data produced by the modern research ecosystem may diminish the value of the metrics in most widespread use today, it also provides an exciting area for future innovation. Scientific communications contain valuable information in the form of text, images, and data files, and are also linked to one another via citations and on social media. At the same time, publishing in web-native formats and in open-access journals (where the results are not locked behind a paywall, and thus are freely available) is increasingly becoming a standard. The combination of these trends makes it possible to develop, track, and potentially even predict, new, nuanced, and targeted metrics of scientific impact.

It is easy to imagine a near future in which scientific impact is quantified by algorithms that, rather than simply counting citations, traverse the full body of scientific literature to extract more nuanced measures. Google's PageRank algorithm, for example, rose to dominance in part because it ranks web pages not by the simple count of references but by weighting each reference by the relative importance of the corresponding web page—references from highly ranked Web sites, like trusted news outlets, therefore score significantly higher than references from lesser-known pages, such as personal blogs. Similar methods can be deployed on the scientific literature and would thus compute not only the number of citations but also the contribution of individual co-authors and the authority of each citing body. This kind of approach, which has recently been shown to most other impact metrics in publication ranking challenges, gives greater weight to citations that come from important or impactful sources, calculated by the same algorithm in a recursive fashion (Brin and Page 1998; Maslov and Redner 2008; Kelly et al. 2018; Xu et al. 2020).

Similar computational methods could then be extended to calculate different dimensions of scientific impact, such as measures of novelty, collaboration, diversity, or interdisciplinarity. Further, as discussed in the next section, the algorithms could be adjusted based on an individual's publication history or stated interests, to suggest the literature, or collaborators, in a tunable way—even recommending the work of highest relevance or optimizing for cross-field insights. In fact, such work is already well underway. Network-based approaches have been used not only to quantify long-term scientific impact but also to identify scientific "gems," measure technological innovation, and quantify the disruptiveness of new work (Wang, Song, and Barabási 2013; Funk and Owen-Smith 2017; Chen et al. 2007). These methods enable more nuanced, granular exploration of the scientific literature and also facilitate new types of research; for example, Wu, Wang, and Evans recently found that there are observable differences in the types of innovations produced by large and small teams, with larger teams tending to build on previous work, and smaller teams more likely to develop disruptive ideas and technologies (Wu, Wang, and Evans 2019).

The new metrics resulting from the application of AI-based methods to the academic publishing ecosystem are transformative. By incorporating more data in the construction of impact scores, and by defining impact in a more nuanced and multi-faceted manner, we will be far more accurate in judging the relevance of research, leading to more efficient hiring and promotion decisions, and thus a more meritocratic scientific ecosystem. Furthermore, the development of these methods will catalyze new research in the science of scientific research and development, with broad implications not only for academic career advancement but also for scientific funding and the optimization of scientific resource allocation more generally.

## SMARTER SEARCH

The changing landscape of academic publishing, and especially the increasing scale and speed of research output, has thrust search and information retrieval into a position of unique importance. Just as increasing scale and complexity in the nascent World Wide Web led to a transition from the Yahoo! index-driven portal page, which was organized around human-curated keywords, to the search-centric Google model, which leveraged computational techniques to identify the most relevant results, so are academic literature searchers increasingly relying on complex information retrieval algorithms. These algorithms currently allow rapid keyword-based retrieval of academic articles, ranked by different parameters. And future models will expand beyond keywords to include smarter metrics and personalized algorithms. In the future, these models could capture enough patterns from the history of scientific research and development to potentially even assist in the creation and assessment of more unique, impactful, and testable scientific hypotheses (Wang et al. 2019).

Currently, many researchers rely on indexes like PubMed and Google Scholar to find articles of interest. However, such methodologies are inexact, and keywords alone are often insufficient to (Gramatica and Pickering 2017) balance the sensitivity and specificity necessary for a successful search. Methods borrowed from branches of AI that deal with both information retrieval and natural language processing (NLP) will allow more granular and customized searches. These algorithms could improve keyword search by, for example, combining it with the searcher's publications, academic co-authorship network, and social media connections, and previous search history to rank results in a personalized manner (Bahmani, Chowdhury, and Goel 2010; Joachims and Radlinski 2007). In combination, topic model-based approaches could use similar papers, rather than keywords, as the inputs to search queries. Companies like Yewno are already exploring the real-world implementation commercialization of such approaches, which augments users ability to navigate search results by providing them with computationally augmented search and filtering mechanisms—for example, allowing search by concept or topic, rather than searching for specific words (Gramatica and Pickering 2017; El-Arini and Guestrin 2011).

Another promising area is the incorporation of computational reasoning into the search process. Future search engines could be designed to return publications most likely to spur disruptive thinking or impactful collaborations, rather than the academic papers most similar to the input keywords. These search algorithms could then be fine-tuned, for example, by broadening or narrowing scope, by filtering out content with specific attributes, or by incorporating work that researchers with similar profiles have found valuable. Even more powerfully, computational parsing of natural language present in scientific articles could be used to auto-construct ontologies of scientific thought which, when used in combination with social networks, would identify the specific parts of research articles that support (or refute) specific queries, and potentially even encourage exposure to alternative viewpoints, account for cultural biases, or identify social dependencies in reasoning (Evans and Rzhetsky 2011; You 2015).

## SMARTER CURATION

The move toward faster publication cycles and increasingly interdisciplinary research puts strain on the time-tested model of academic peer review prior to publication. At the same time, the growing adoption of preprint servers like arXiv and bioRxiv better aligns with a post-publication peer review model, which many have argued results in more efficient and less bias-prone curation (Sullivan 2018). While some fields, like mathematics and physics, have relied on public discourse around research findings for some time, other disciplines are finding themselves thrust into a form of online, crowdsourced peer review spanning many sites and applications, from blog posts to Twitter conversations (Mandavilli 2011).

AI-based methods can be used to help organize, structure, and extract value from the many diverse conversations circulating online around academic research. They can also help us make the best possible use of high-quality peer review, which should be considered a sparse, valuable resource. For example, sentiment analysis, a method from NLP that allows the positive or negative valence of a comment to be quantified, could be used to extract more nuanced meaning from Tweets, blogs, and comments. Then, prediction models that take the quantity and content of online conversations into account could be used to determine when a preprint article has garnered sufficient positive attention to merit peer review. Learning algorithms could even be used to quantify the quality of reviews, so that the peer reviewers with the best "track record" or history of correctly identifying positive or negative indicators in specific fields are allocated to the papers in most need of their skills.

Finally, AI algorithms can also be deployed to increase reproducibility in science. Existing NLP methods are increasingly being leveraged to compare texts for similarities with proceeding *work* in an author-specific manner, and further research in this vein is current and ongoing (Soleman and Purwarianti 2014; Bandara and Wijayarathna 2011; Burrows, Potthast, and Stein 2013). Machine vision algorithms, including convolutional neural networks (CNNs), could be used to detect image and figure reuse at broad scale (Acuna, Brookes, and Kording 2018), and network-based approaches could be leveraged in combination with large databases of retracted papers, such as *Retraction Watch Database*, to identify the warning signs of irreproducibility (Retraction Watch Database 2019).

## CONCLUSION

The application of artificial intelligence and machine learning methods to the academic research and communication process holds significant promise— not only to help us make sense of the existing literature landscape but also to help us generate breakthrough insights and technologies more rapidly and efficiently. In this brief chapter, we have described a subset of these application domains in order to illustrate this exciting potential.

## REFERENCES

Acuna, D. E., Brookes P. S., and Kording K. P. February 23, 2018. "Bioscience-Scale Automated Detection of Figure Element Reuse" [Internet]. doi:10.1101/269415

Bahmani, B., Chowdhury, A., and Goel, A. 2010. "Fast Incremental and Personalized PageRank" [Internet]. *Proceedings of the VLDB Endowment.* 4 (3): 173–184. doi:10.14778/1929861.1929864

Bandara, U., and Wijayarathna, G. 2011. "A Machine Learning Based Tool for Source Code Plagiarism Detection" [Internet]. *International Journal of Machine Learning and Computing.* 1 (4): 337–343. doi:10.7763/ijmlc.2011.v1.50

Brin, S., and Page, L. 1998. "The Anatomy of a Large-Scale Hypertextual Web Search Engine. Computer Networks and ISDN Systems." *Elsevier* 30: 107–117.

Burrows, S., Potthast, M., and Stein, B. 2013. "Paraphrase Acquisition via Crowdsourcing and Machine Learning" [Internet]. *ACM Transactions on Intelligent Systems and Technology.* 4 (3): 1. doi:10.1145/2483669.2483676

Chen, P., Xie, H., Maslov, S., and Redner S. 2007. "Finding Scientific Gems with Google's PageRank Algorithm." *Journal of Informetrics.* 1: 8–15.

Csiszar, A. 2017. "How Lives Became Lists and Scientific Papers Became Data: Cataloguing Authorship during the Nineteenth Century—Corrigendum." *British Journal for the History of Science.* 50: 567.

El-Arini, K., and Guestrin, C. 2011. "Beyond Keyword Search [Internet]." *Proceedings of the 17th ACM SIGKDD International Conference on Knowledge Discovery and Data Mining—KDD'11.* doi:10.1145/2020408.2020479

Evans, J. A., and Rzhetsky, A. 2011. Advancing Science through Mining Libraries, Ontologies, and Communities. *Journal of Biological Chemistry.* 286: 23659–23666.

Franck, G. 1999. "Essays on Science and Society: Scientific Communication—A Vanity Fair?" *Science* 286: 53–55.

Funk, R. J., and Owen-Smith J. 2017. "A Dynamic Network Measure of Technological Change" [Internet]. *Management Science.* 63 (3): 791–817. doi:10.1287/mnsc.2015.2366

Garfield, E. 1955. "Citation Indexes for Science A New Dimension in Documentation through Association of Ideas." *Science* 122: 108–111.

Garfield, E. 2006. "The History and Meaning of the Journal Impact Factor." *JAMA* 295: 90–93.

Gramatica, R., and Pickering, R. 2017. "Start-up Story: Yewno: An AI-driven Path to a Knowledge-Based Future" [Internet]. *Insights: the UKSG Journal.* 30 (2): 107–111. doi:10.1629/uksg.369

Hirsch, J. E. 2005. "An Index to Quantify an Individual's Scientific Research Output." *Proceedings of the National Academy of Sciences of the United States of America* 102: 16569–16572.

Joachims, T., and Radlinski, F. 2007. "Search Engines that Learn from Implicit Feedback" [Internet]. *Computer.* 40 (8): 34–40. doi:10.1109/mc.2007.289

Kelly, B., Papanikolaou, D., Seru, A., and Taddy, M. 2018. "Measuring Technological Innovation over the Long Run" [Internet]. doi:10.3386/w25266

Mandavilli, A. 2011. "Peer Review: Trial by Twitter." *Nature.* 469: 286–287.

Maslov, S., and Redner, S. 2008. "Promise and Pitfalls of Extending Google's PageRank Algorithm to Citation Networks." *Journal of Neuroscience* 28: 11103–11105.

Retraction Watch Database [Internet]. [cited July 29, 2019]. Available at: http://retraction
    database.org/RetractionSearch.aspx?
Soleman, S., and Purwarianti, A. 2014. "Experiments on the Indonesian Plagiarism
    Detection Using Latent Semantic Analysis" [Internet]. *2014 2nd International
    Conference on Information and Communication Technology (ICoICT)*. doi:10.1109/
    icoict.2014.6914098
Sullivan, Bill. 2018. "Is It Time for Pre-Publication Peer Review to Die?" [Internet]. *PLOS*.
    [cited August 9, 2019]. Available at: https://blogs.plos.org/scicomm/2018/08/28/
    is-it-time-for-pre-publication-peer-review-to-die/
Wang, D., Song, C., and Barabási A.-L. 2013. "Quantifying Long-Term Scientific Impact"
    [Internet]. *Science*. 342 (6154): 127–132. doi:10.1126/science.1237825
Wang, Qingyun, Huang, Lifu, Jiang, Zhiying, Knight, Kevin, Ji, Heng, Bansal, Mohit, and
    Luan, Yi. 2019. "PaperRobot: Incremental Draft Generation of Scientific Ideas" [Inter-
    net]. [cited July 29, 2019]. Available at: https://arxiv.org/pdf/1905.07870.pdf
Wilhite, A. W., and Fong, E. A. 2012. "Scientific Publications. Coercive Citation in Aca-
    demic Publishing." *Science* 335: 542–543.
Wu, L., Wang, D., and Evans, J. A. 2019. "Large Teams Develop and Small Teams Disrupt
    Science and Technology." *Nature* 566: 378–382.
Xu, S., Mariani, M. S., Lü, L., and Medo, M. 2020. "Unbiased Evaluation of Ranking Met-
    rics Reveals Consistent Performance in Science and Technology Citation Data." *Journal
    of Informetrics* 14 (1).
You, J. 2015. "Artificial Intelligence. DARPA Sets Out to Automate Research." *Science*
    347: 465.

# CHAPTER 4

# KNOWLEDGE DISCOVERY IN THE AI ERA: KNOWLEDGE GRAPHS, A NEW DATA FRAMEWORK TO TACKLE UNSTRUCTURED EVER-GROWING INFORMATION

Ruggero Gramatica, PhD

Back in 2010, I was chatting with a gentleman on an airplane about how the information economy has become, in the past 50 years, one of the most rapidly growing industries in the world. There has been a proliferation of companies over this period with the ability to produce information, make information available, and create infrastructure (e.g., databases and applications) serving the purpose of structuring information to be extracted and analyzed. This, in turn, has catalyzed a shocking increase of data production.

The amount of information presently produced has reached a level that would have been unimaginable only a couple of decades ago. In 2018 the global data supply reached 33 zettabytes (ZB)—or 33 trillion GB—though, despite the value of this data, just less than 5 percent of it is used for analysis. Volumes of data are projected to reach 125 ZB by 2025, with emerging economies accounting for an increasingly large proportion of the world's total (Reinsel, Gantz, and Rydning 2018).[1]

The good news for those who use data is that someone somewhere is producing the information we need. The bad news is that such information is fragmented and dispersed, and we do not know how to structure it for use. Moreover, no matter how much information is available, information is not the same as knowledge. We may be able to access a large chunk of available information, but that chunk of data does not make us any more

---

1 David Reinsel, John Gantz, and John Rydning. November 2018. "The Digitization of the World: From Edge to Core." *IDC Whitepaper #US44413318*. Sponsored by Seagate.

knowledgeable about a specific subject. In fact, information must be read, interpreted, and cognitively understood in order for us to act upon it.

But those in the knowledge economy need to deal with more than just the ever-increasing amount of data. The key challenge for those working with large amounts of data is, in fact, the intrinsic complexity of that data. It is the unstructured complexity of data that makes the extraction of hidden information out of the mass of data challenging. Data that are both unstructured and ever-increasing are very difficult to analyze.

## THE EMERGENCE OF THE KNOWLEDGE GRAPH AND THE KNOWLEDGE ECONOMY

Fortunately, recent developments in parallel processing, cloud storage, and computing have facilitated the development of machine learning technologies, including today's blooming of artificial intelligence (AI), which require huge amounts of data to train the algorithms that drive them. The availability of these huge data sets has allowed us to leave behind the information economy and enter a new era: *the knowledge economy.*

There are many common tools and modern business intelligence methods that allow analysts to delve into complex datasets and provide general answers to predetermined questions. That is, we can search databases using keywords and draw information out of them fairly effectively. However, it is undoubtedly also true that when we use such unstructured data we often come to inconclusive results, and the potentially useful information remains hidden from us.

There are now innovative, new AI-based tools that will help us to extract more information from our data. Perhaps the most promising new technology is called *the Knowledge Graph Data Framework*, a methodology we can use to construct a graph-based knowledge framework derived from semantically processing large corpora of unstructured data. The graph-theoretic approach helps us to resolve the intrinsic complexity of unstructured data and to extract meaningful insights hidden in the data (emergent properties), and then to develop tools that highlight the meaningful information (Treur 2020).[2]

---

2 Treur, J. (2020). "Relating Emerging Network Behaviour to Network Structure." In: *Network-Oriented Modeling for Adaptive Networks: Designing Higher-Order Adaptive Biological, Mental and Social Network Models. Studies in Systems, Decision and Control*, vol. 251. Cham: Springer. https://doi.org/10.1007/978-3-030-31445-3_11.

Tools and infrastructure for graph-based knowledge representation are coming to market to help professionals in multiple fields analyze massive data sets in order to find non-obvious connections that can lead to better decisions, providing the ability to find hidden relationships that can lead to profitable discoveries while increasing researcher and analyst productivity (Van den Broeck 1991).[3]

Knowledge Graphs—one of the most promising AI-based developments—recently made their appearance on the Gartner Hype Cycle, which charts the development and adoption of new technologies. They appear in the Innovation Trigger area of the cycle, with a 5-to-10 year trigger for mass adoption.

When thinking of the power of Knowledge Graphs, we can imagine an ever-growing stream of unstructured data flowing into a repository in which the data aggregates formlessly. The process recalls a complex physical system in which parts of the system are governed by certain local and global properties and dynamics. In this context, the study of complex systems comes in handy in extracting *emergent properties*, which become key in the interpretation of knowledge (Brede 2012).[4]

Emergent properties are attributes of the "whole" (in our example, the gigantic repository of information) that are not strictly driven by any of the individual parts making up that whole. Such phenomena exist in various domains and can be described using complexity concepts and thematic knowledge (Aziz-Alaoui and Bertelle 2009).[5]

The term "knowledge graph" may have emerged into the business mainstream through Google's use of the term to describe their vast web data representation that powers its search engine functionality. But this doesn't quite capture what knowledge graphs are (Chen 2019).[6]

---

3 Van den Broeck, C. (1991). "Entropy and Learning." In: Babloyantz, A. (ed.) *Self-Organization, Emerging Properties, and Learning. NATO ASI Series (Series B: Physics)*, vol. 260. Boston, MA: Springer. https://doi.org/10.1007/978-1-4615-3778-6_16.

4 Brede, Markus. 2012. "Book Review: *Networks—An Introduction* by Mark E. J. Newman (2010, Oxford University Press.) $65.38, £35.96 (hardcover), 772 pages. ISBN-978-0-19-920665-0." *Artificial Life* 18 (2): 241–242. doi: https://doi.org/10.1162/ARTL_r_00062.

5 Aziz-Alaoui, M. A., and Bertelle, C., eds. 2009. *From System Complexity to Emergent Properties*. Understanding Complex Systems Series. doi:10.1007/978-3-642-02199-2.

6 Chen, Xiaowei Luo, 2019. "An Automatic Literature Knowledge Graph and Reasoning Network Modeling Framework Based on Ontology and Natural Language Processing." *Advanced Engineering Informatics* 42, 100959, ISSN 1474–0346, https://doi.org/10.1016/j.aei.2019.100959.

A true knowledge graph correlates unstructured information in a multi-dimensional way over time (Gramatica 2019).[7]

A full-fledged knowledge graph can find correlations across essentially any type of data set, doing the kind of inferential work it might take a full team of analysts with access to large information data sets many hours to perform. Used properly, a knowledge graph provides a framework to *extract knowledge*, that is, inferential pathways connecting several objects, taking points of *information* and turning them into usable *knowledge* (Fensel, et al. 2020).[8]

## A RELATIONSHIP-DRIVEN FRAMEWORK

Extracting knowledge from complex and articulated data inputs requires a flexible structure; a data framework organized as a graph (hence a knowledge graph) provides an ideal staging structure for a dynamic ontology layer and other type of relational functions that define the set of types, properties, and relationships continuously adjusting as new information comes through. While other graphs map a picture of predefined relationships between entities, those tend to be limited, as they can't easily account for the introduction of "inferred" data from other emerging relationships (Zou 2020).[9]

Imagine, for example, the relationships that can be dynamically extracted when tracking business interactions between companies. If you have access to certain private databases, you can have a supply chain link between the two companies. But such a relationship might not be enough; it is important to understand, in addition to all the factual types of relationships the two entities have in common (i.e., sector, investor relationships, size/market commonalities, affiliation, partnerships, and so on), that others can be inferred by second- or third-order connections. For example, if both entities are exposed to political, financial, or geographic risks, a link with such emerging relationships will enrich other, already existing ones. That is what a knowledge graph with a dynamic relationship matrix looks like.

7  Gramatica, Ruggero, 2019. *Going Under the AI Hood: What Knowledge Graphs Really Are and How They Focus Decision Making*. LinkedIn July 2, 2019. https://www.linkedin.com/pulse/going-under-ai-hood-what-knowledge-graphs-really-how-gramatica-phd/

8  Fensel, D., et al. 2020. "Introduction: What Is a Knowledge Graph?" In: *Knowledge Graphs*. Cham: Springer. https://doi.org/10.1007/978-3-030-37439-6_1.

9  Zou, Xiaohan, 2020. "A Survey on Application of Knowledge Graph." *Journal of Physics: Conference Series*. 1487 012016.

In other words, the amount of information that gravitates directly or indirectly around any concept generates in this type of AI-based data framework an induced, ever-changing layer of relationships that dramatically enriches the quantum of information between any entities of the graph.

Knowledge graphs are also actual graphs—based on mathematical models—which make them easily expandable and allow for the application of mathematical graph theory techniques, such as network analysis. That ability to process and interpret plain language is at the cornerstone of what makes them unique.

## AN EXPANDABLE LEARNING MODEL

Ultimately, an AI-based data framework formalizes and standardizes what has traditionally been a uniquely human problem: the transformation of unstructured and diverse types of information into knowledge (hence the term "knowledge graph")—a framework that can be sliced and diced, unfolded even over hundreds of millions of individual objects and billions of relationships.

Intelligent machine learning algorithms refine their logic as data sets are introduced, which means knowledge graphs, unlike traditional analysis, actually perform better and more efficiently as the number of information constituents grows. Doing this properly requires far more than a database: components such as hierarchical topic models, dynamic taxonomies and ontologies, and inference engines are all necessary to ensure a sustainably developed knowledge graph (Liu and Han 2018).[10]

The goal in all of this is simply to uncover the intimate relationships among chains of concepts—across domains—in an evolving and expandable corpus of data. For example, a knowledge graph does not only need to know that "Lehman Brothers" refers to a U.S. investment bank, but in addition that in 2008, "Lehman Brothers" suddenly became strongly linked and exposed to concepts like "global crisis," "bankruptcy," and "subprime mortgage crisis," thus reshaping its inferential meaning. These analytical concepts can then be linked together with other inferences extracted in a similar way, permitting the creation of a socioeconomic cluster leading to further analysis, prediction, and generation of hypotheses.

Like any evolving technology, knowledge graphs have developed into something of a buzzword that will continue to obfuscate their meaning for

---

10  Liu, Z., and Han, X. 2018. "Deep Learning in Knowledge Graph." In: Deng, L., and Liu, Y. (eds.) *Deep Learning in Natural Language Processing*. Singapore: Springer. https://doi.org/10.1007/978-981-10-5209-5_5.

some time to come, with a few players stepping in to claim their own graphs. The real—and most valuable——uses of the technology, though, will be inferentially driven, expanding data sets to derive correlations on a scale that would have been completely impossible for human hands alone.

That level of analysis offers real knowledge on a scale that would have been unthinkable under more traditional analyst-driven models, which can only enhance the level of information on which people are capable of acting. And it will apply to multiple fields of human knowledge, from social science, pharmacology, finance, history, politics, and beyond.

Welcome to the Knowledge Economy!

## REFERENCES

Aziz-Alaoui, Moulay, and Cyrille Bertelle, eds. From System Complexity to Emergent Properties. *Understanding Complex Systems Series*. Berlin: Springer, 2013.

Babloyantz, Agnessa, and Christian Van den Broeck. "Entropy and Learning." Essay. In *Self-Organization, Emerging Properties, and Learning* 260, 260:231–40. NATO ASI Series. New York, NY: Plenum Press, 1991.

Brede, Markus. "Book Review: Networks-An Introduction by Mark E. J. Newman." *Artificial Life* 18, no. 2 (February 2012): 241–42. https://doi.org/10.1162/artl_r_00062.

Chen, Hainan, and Xiaowei Luo. "An Automatic Literature Knowledge Graph and Reasoning Network Modeling Framework Based on Ontology and Natural Language Processing." *Advanced Engineering Informatics* 42 (October 2019): 100959. https://doi.org/10.1016/j.aei.2019.100959.

Deng, Li, and Yang Liu. "Deep Learning in Knowledge Graph." Essay. *In Deep Learning in Natural Language Processing*, 117–45. Puchong, Selangor D.E: Springer Singapore, 2018.

Fensel, Dieter, Şimşek Umutcan, Kevin Angele, Elwin Huaman, Kärle Kevin, Oleksandra Panasiuk, Ioan Toma, Umbrich Jürgen, and Alexander Wahler. "What Is a Knowledge Graph?" Introduction. In *Knowledge Graphs: Methodology, Tools and Selected Use Cases*, 1–10. Cham: Springer, 2020.

Gramatica, Ruggero. "Going Under the AI Hood: What Knowledge Graphs Really Are and How They Focus Decision Making." *LinkedIn*. LinkedIn, July 2, 2019. https://www.linkedin.com/in/ruggero-gramatica-phd-1395054.

Reinsel, David, John Gantz, and John Rydning. "The Digitization of the World from Edge to Core." *International Data Corporation*. Seagate, November 2018. https://www.seagate.com/files/www-content/our-story/trends/files/idc-seagate-dataage-whitepaper.pdf.

Treur, Jan. "Relating Emerging Network Behaviour to Network Structure." *Network-Oriented Modeling for Adaptive Networks: Designing Higher-Order Adaptive Biological, Mental and Social Network Models*, 2019, 251–80. https://doi.org/10.1007/978-3-030-31445-3_11.

Zou, Xiaohan. "A Survey on Application of Knowledge Graph." *Journal of Physics: Conference Series* 1487, no. 1 (2020): 012016. https://doi.org/10.1088/1742-6596/1487/1/012016.

# CHAPTER 5

# CONCEPTS, NOT KEYWORDS: SEARCH MADE INTELLIGENT
Haris Dindo

## INTRODUCTION

The experience of information overload has become a serious challenge for STEM researchers in the era of networked information. The enormous quantity of online data and information made public by researchers, institutions, and governments means that researchers face the daunting task not only of discovering widely scattered information but also of determining the relevance of what they have discovered to their own projects and decision-making.

In fact, it is not only researchers who are experiencing information overload: traditional search and discovery tools are also overwhelmed as they seek to discover and process what is relevant from the mountain of information available online. To help researchers to deal with information overload, we need more advanced tools that not only help them to discover the relevant information but also to disregard the information that is not relevant.

We can describe the mass of information available to us as complex systems— systems characterized by an immense number of elements that are neither homogeneous nor linear. Examples of complex systems are ant colonies, the human brain, opinion dynamics, massive data sets, and financial markets, all of which require a research approach that accepts that the relation between the parts of the system cannot be reduced to a simple model in which one plus one equals two.

Rather, complex systems oscillate back and forth between randomness and order, the behavior of the ant hive reflecting both the orderly functioning of instinct and hierarchy and the disorderly response to disruption and changing circumstance, whether the arrival of an anteater or a flood that fills the hive.

Distinguishing instances of order and regularity within a complex system is one of the chief challenges of the science of complexity: once the correct level of description is found, systems that previously appeared random and unpredictable can be (at least temporarily) controlled, monitored, and predicted.

How can a researcher search massive and differentiated networks of information so that the output of their search is neither overwhelming nor littered with irrelevant results? To improve discovery, we need to go beyond the current discovery tools that search for keywords and provide only the most basic ranking of the output. Rather, we need discovery engines that can analyze complex masses of information and use new artificial intelligence (AI)–based models to extract the information we need from them.

Rather than have the user sift through the output of a search to extract what is relevant—reading abstracts or skimming the articles—we can, with a new kind of discovery tool, convert enormous volumes of unstructured data into usable outputs, which are intuitively packaged for the user. Bundles of heterogeneous information and data are translated by the discovery tool into interconnected concepts and layered one on top of another, providing different points of view on the underlying phenomena. The output of these new tools is not a single concept or aggregated lists of concepts, which ignores the complexity of the system within which the data exist. These new discovery tools, instead, focus on the behavior of a *network of interconnected concepts*. Concepts could be interconnected because they are semantically similar, because they co-occur frequently together, because there is a factual relationship between them, or any combination of the above. This approach also explicitly addresses the temporal evolution of the network by tracking changes in these connections over time. Collectively, said networks are called a knowledge graph (KG).

## CONCEPTS, NOT KEYWORDS

Human language is both complex and nuanced, and this is equally true of masses of unstructured data. If we wish to extract relevant knowledge from this data, we need to go beyond traditional information retrieval tools that focus upon the presence of certain *keywords* in the corpus of data and that currently permeate our interaction with the technology (think of spam filters, spell checkers, translation tools, etc.). Approaches that operate on keywords directly suffer from two problems: *polysemy* (words might have multiple meanings, think of the word "bank") as well as *synonymy* (different words might have the same meaning: USA and US refer to the same entity, United

States of America). Both issues can be overcome by adopting *concepts* as basic information units as further elaborated in the following.

A concept is defined as a set of strongly related words, *synonyms*, gathered together under the same topic. The word "bank," for instance, can either be the place where we store our money or the borders of a river. This is an example of *polysemy*, which is inherent to the ambiguous nature of the human language. How do we know when we are referring to the financial institutions rather than to the geographic features of a continent? We can use embedded identifiers, such as topics [topic: finance] or [topic: geography], to establish the semantic value of a word and its subsequent relevance to a concept.

Each concept in similarly designed system represents a unit of knowledge with a number of different qualities: a well-defined semantics, a set of associated topics, a uniquely defined identifier, one or more definitions describing what that concepts are about, and a set of synonyms preferentially used to refer to the concept in the unstructured data.

Figure 5.1 shows the concept "automobile," characterized by its (potentially multiple) definitions, set of terms (synonyms), and a topic of transportation. All concepts are collectively stored into an expandable knowledge base.

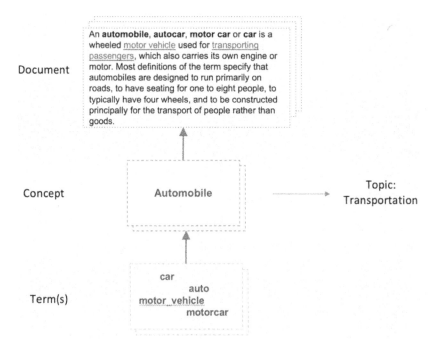

**Figure 5.1. Concepts in a Knowledge Base.**

Having defined a knowledge base of concepts, how do we connect related concepts together in a purely data-driven fashion (and without any human intervention)? Recent advances in AI allow us to mathematically capture the *semantics* of any portion of text through a mathematical operation of *embedding*. Without getting into mathematical details, the idea is that concept definitions are projected into a metric space so that semantically similar concepts are topologically close. However, not only concept definitions are embedded: any portion of text (e.g., a paragraph from a book) can be projected onto the same space and compared to other projections (i.e., other paragraphs of text or definitions of concepts). This brings us to a new generation of search engines, whereby search can be performed not by matching keywords along predefined rules but by extracting semantically similar portions of text to a given query. In addition, this very same process of comparing concepts and portions of text where they potentially appear can be adopted for concept extraction and disambiguation, essential components for building the knowledge graph, which interconnects the concepts based on different similarity criteria.

## KNOWLEDGE GRAPH

The goal of the knowledge graph is to uncover the intimate relations between various concepts from a timely and evolving corpus of data. In our view, it is the interaction between concepts—expressed as a network of relations at a point in time—that binds (temporarily) a meaning to the involved concepts.

A link between any two concepts in the knowledge graph, furthermore, is induced by their similarity along several independent dimensions:

*Semantic:* Concepts that occur in similar contexts share semantic properties and are connected in the graph.

*Syntactical:* Concepts that are found in a strong subject-verb-object (SVO) relationship are connected through what we call an inferential network. Not only are the connections created, but since those connections are extracted from factual examples, these can be explained using plain text. For instance, the sentence "An antibiotic is a type of antimicrobial substance active against bacteria" indicates a strong SVO relationship between "antibiotic" (a concept) and "bacteria" (another concept), via the verb (also called *predicate*) "active against."

*Exposures:* The similarity measures discussed here implicitly induce first-order graphs, whereby similar concepts are directly connected, as in the previous example between antibiotic and bacteria. However, it is possible to use these graphs to extract second-order, or *indirect*, similarity measures. Exposure-based similarity is one such measure, where we estimate the potential influences between any pair of nodes in the network, irrespective of their direct or indirect connection. Such an algorithm might, for instance, uncover a potential exposure between the antibiotic and another disease through a series of indirect links, enabling de facto a serendipitous discovery of new ideas.

## SCIENTIFIC FOUNDATIONS AND CONCLUSION

When we build systems to process unstructured data, our first difficulty is that we are dealing primarily with words as the "observable" counterparts of concepts, and thus we face the problem of semantics (polysemy and synonymy). A viable approach to extract the *meaning from text can be built by expanding* upon the so-called distributional hypothesis, according to which words with similar meanings occur in similar contexts. In other terms, there is a tight link between *distributional similarity* and *meaning similarity*, which allows us to use the former in order to estimate the latter (Sahlgren 2008). A similar approach has been adopted successfully, for instance, in the biomedical field, with the goal to uncover relations between medical entities (Gramatica et al. 2014). However, similar co-occurrence–based methods—the most common way of estimating distributional properties from data—fall short when it comes to polysemy and synonymy.

Approximate distributional hypotheses can be approximated via a neural network–based structure trained on the "local context" of each word. This approach effectively translates words (and their multiple meanings) into a vector space of fixed cardinality, wherein "semantically similar" words are closely grouped together.

In contrast to these approaches, however, we propose that the points in a semantic space are well-formed *concepts* rather than words. In this respect, our approach shares a number of similarities with Gärdenfors's theory of conceptual spaces (2014). Indeed, our space accounts for geometrical/topological operations (e.g., similarity computation, clustering, traversing) and enables the adoption of "quality dimensions" that are humanly readable. This is especially useful when assessing the relations between concepts, yielding to diverse levels of similarity corresponding to multiple partitions of the conceptual space. (As an example, the concepts "Merlot" and "Syrah" can be strongly related via distributional hypothesis over a given corpus but also via additional quality dimensions describing, e.g., more abstract relationships, such as the one of both being wines, both having a similar color—in this case, red—or other types of mutual interactions and dependencies).

Similarity between concepts implicitly induces a graph-like network permitting even more subtle analysis by adopting tools from the realm of complex networks (Newman 2003). The geometrical notion of *closeness* between any two concepts is translated into the existence of a direct link between the same concepts in the corresponding network. Moreover, similarity along different quality dimensions can be represented via a multilayered network (multiplex), where different layers are mapped to different partitions of the corresponding conceptual space (Scott 2000).

So far, we have focused on the static data. Continuous ingestion of data adds an additional dimension: time. More precisely, continuous data ingestion explicitly addresses the temporal evolution of the conceptual networks described earlier and extracts insights by analyzing not only the nature of the interconnected concepts but also their temporal evolution. This allows the evolution of semantics to be quantified and establishes a link to other closely related real-world phenomena that might be influenced, either directly or indirectly, by said dynamics. (e.g., it is well known that information contained in news might have a direct effect on stock returns.)

Finally, our framework is in line with certain theories from the cognitive brain sciences postulating that the organization of conceptual knowledge in the brain reflects the statistical co-occurrence of certain observable properties (Caramazza and Mahon 2006). This so-called correlated structured principle postulates that conceptual features corresponding to properties that often co-occur will be stored close together in semantic space. At the neuronal level, this implies that conceptual knowledge corresponding to objects with similar properties is stored in adjacent neural areas, forming a network of interrelated concepts. Focal brain damage can give rise to category-specific semantic deficits as damage to a given property will propagate damage to highly correlated properties. We conjecture that the effects of such phenomena in the brain could be simulated through large-scale "diffusion" processes on our conceptual network, thus studying the global effects of local "perturbations" at the level of concept connectivity (see also Abdelnour, Voss, and Raj 2014 on modeling the relationship between structural and functional patterns in the brain via network diffusion). We foresee that similar efforts could also provide a novel insight into the biological mechanisms of certain brain disorders.

## REFERENCES

Abdelnour, F., Voss, H. U., and Raj, A. 2014. "Network Diffusion Accurately Models the Relationship between Structural and Functional Brain Connectivity Networks. *Neuro-Image* 90: 335–347.

Caramazza, A., and Mahon, B. Z. 2006. "The Organisation of Conceptual Knowledge in the Brain: The Future's Past and Some Future Directions." *Cognitive Neuropsychology* 23 (1): 13–38.

Gärdenfors, P. 2014. *The Geometry of Meaning: Semantics Based on Conceptual Spaces*. The MIT Press. Cambridge; London.

Gramatica, R., Matteo, T. D., Giorgetti, S., Barbiani, M., Bevec, D., and Aste, T. 2014. Graph Theory Enables Drug Repurposing—How a Mathematical Model Can Drive the Discovery of Hidden Mechanisms of Action. *PLOS ONE* 9 (1): 84912.

Newman, M. E. J. 2003. "The Structure and Function of Complex Networks." *Siam Review* 45 (2): 167–256.

Sahlgren, M. 2008. "The Distributional Hypothesis." *The Italian Journal of Linguistics* 20 (1): 33–54.

Scott, J. 2000. *Social Network Analysis*. SAGE Publications Ltd. Thousand Oaks, CA.

# CHAPTER 6

# WOULD A GOOGLE OF AIS BE ABLE TO PREDICT THE FUTURE?: WHERE IS INFORMATION MANAGEMENT HEADED IN A WORLD OF ARTIFICIAL INTELLIGENCE?

*Todd A. Carpenter*

Over the past 60 years, computers have transformed how content is created, organized, retrieved, and even how it is consumed. While most of us have become comfortable managing content with computers, the growth of artificial intelligence (AI) is affecting not only how we engage with content but how content is created, navigated, and understood. This is an entirely new way for us to rely on technology, and while we sometimes assume that advances in AI can lead to the displacement of humans in knowledge work, this need not be the case.

Before we can understand the future of *machine* intelligence, we should certainly ask if we even have a common understanding or expectation of what *human* intelligence is. If one were to ask a random assortment of passersby to define intelligence, one suspects that there would be no common definition among them. And, indeed, we all know that various people have very different intellectual strengths. One person who is brilliant at mathematics or physics, after all, might be quite clueless in the world of practical affairs. The brilliant orator might not be able to balance his or her checkbook. An athlete able to read an opponent's physical "tells" and therefore win championships but might not be able to speak eloquently. The most artistic pianist might not be able to remember where they left their car keys.

The function and mystery of the mind has occupied the attention of countless philosophers, neurologists, psychologists, and linguists over many centuries. Tens of thousands of books have been written about the human mind and how it works. Given how intimately connected we are to our minds, the most amazing thing is that our minds are still poorly understood. And if our own minds remain so mysterious, it is not surprising that we have had limited success translating that vague understanding we do have into silicon, transistors, and computational switches.

To be sure, narrow AI—that is, the application of AI toward single tasks or problem-solving—has made great strides in recent years. But this narrow form of machine intelligence does not necessarily translate to intelligence in other spheres—something that applies equally to humans as it does to machines. There may only be a few instances of machine learning that can perform in multiple spheres, but the true polymath is also a very rare human being.

One of the errors in expecting the creation of a human-like AI is that we commonly underestimate the underlying complexity of many basic human activities. We mistake the narrow capacity of a machine in a very specific task or domain for a more general intelligence. There is not now, nor will there be any time soon, a generalized AI that is as adaptable as even a chimpanzee or a dog. Machines can be good—far better than humans in some cases—at a very limited set of activities that require rapid computation or the recognition of patterns in data. Some of the things machines do might even fall under the rubric of intelligent behavior. Yet "intelligence" is the wrong word for what a machine does in a very narrow context.

Given the everyday examples of how AI is used today, from face recognition software to voice recognition, it is well to remember that machines were, until recently, incapable of any sort of image recognition, nor could they recognize and transcribe human speech. Today image and voice recognition technologies are commonplace, if not ubiquitous. With each such advance in machine information processing, the realm of "magic"—in the context of some kind of interaction that is beyond normal belief—is pushed further and further into what had been science fiction.

While we may marvel at the accomplishments of machines, it is important to remember that current computational technology is still based on switches. Machine languages are fundamentally based on zeros and ones; essentially grounded in a "yes" or "no" logic. Given this positive or negative foundation, machine learning has had to resort to statistical probabilities to address the world's complexities.

For example, while we may be impressed with the ability of our phone to auto-complete our sentences as we type, we should understand that in language, given the limited number of words and the rules of grammar and syntax, it is possible to compute with some level of confidence what the next word will be. The more one strays from the ordinary, say in more creative works, the more this system begins to break down. For example, R. R. Martin wrote his *Game of Thrones* series using an old word processor because newer "more intelligent" systems were constantly trying to fix his "errors" in words and syntax.[1] It is in the departure from statistics and probability toward higher-level intellectual domains, such as wisdom and creativity, that the boundaries of AI systems lie.

In the world of research, discovery, and scholarly communication, AI will have an impact in five key areas: storage and retrieval, machine reading, understanding, synthesis, and creativity. Much like the hierarchy of knowledge,[2] moving from memorization to wisdom, AI will play a role in all these layers moving into the future, though with diminishing contributions up the value chain of human creativity. The value added differs at each level up the chain and the time horizon for advances extend outward into the future for each level, as the issues grow in complexity with each new level.

## STORAGE AND RETRIEVAL

Training an AI system is a data-intensive enterprise. In a probabilistic approach to problem-solving, the more data upon which to test, process, and compare the AI's results, the more likely it is that the output will improve. Extending beyond simple gathering and storage of the training data, a probabilistic approach to analysis requires enough memory to store a stepwise approach to

1 Martin, George R. R. May 14, 2014. "Why I Still Use DOS." Website: https://www.bbc.com/news/technology-27407502 (accessed September 20, 2020).
2 Frické, Martin, 2019. "The Knowledge Pyramid: The DIKW Hierarchy." *Knowledge Organization* 49 (1): 33–46. Also available in *ISKO Encyclopedia of Knowledge Organization*, eds. Birger Hjørland and Claudio Gnoli, http://www.isko.org/cyclo/dikw

problem-solving. One reason for this is because the system needs to maintain consistency as it works through the problem. The lengthier or more complex the problem, the more challenging the issue of memory becomes for the system. To take an example, if one is writing a paragraph, remaining consistent over the space of a couple hundred words is one thing; consistency over hundreds of pages is more difficult. Both collecting and storing training data and processing memory require significant amounts of storage and data retrieval capacity to address these complex questions.

Although it is less well known than Moore's Law,[3] there is a similar principle related to the density of storage,[4] whereby the cost and memory capacity of hard disk storage space decrease at similar logarithmic pace as computational power of microprocessors have increased over time. Disk storage technology may be potentially approaching the physical limits of existing storage mediums, as noted by David Rosenthal and others in the mid-2010s[5] that actual production was beginning to diverge from Kryder's Law projections. However, there are other approaches that could possibly continue the explosive growth in storage capacity. For example, the issue of capacity could be solved not by individual devices becoming more and more powerful, but by networking devices at scale, an approach that is widely adopted across the Internet, although this approach compounds the problem of data retrieval, which will be discussed later. Beyond this on the horizon, there are other potential mediums for storage, such as biological structures[6] that could continue to push the boundaries of storage capacity. Realistically, despite the explosion of the need for additional storage, this issue is most likely among those related to AI to be resolved and therefore is not likely to be a hindrance to the advancement of either narrowly defined or more complex AI. Simply having the information to hand, being able to retrieve it electronically, and having the storage to manage all those data are a prerequisite for higher-level analysis of the information.

---

3 Moore's Law. Wikipedia. https://en.wikipedia.org/wiki/Moore%27s_law
4 Kryder's Law. https://en.wikipedia.org/wiki/Mark_Kryder#Kryder's_law_projection
5 Rosenthal, David. December 13, 2016. "The Medium-Term Prospects for Long-Term Storage Systems." https://blog.dshr.org/2016/12/the-medium-term-prospects-for-long-term.html
6 Gent, Edd. April 1, 2019. "Microsoft Is Building an All-in-One DNA Data Storage Device." Website: https://singularityhub.com/2019/04/01/microsoft-is-building-an-all-in-one-dna-data-storage-device/ (accessed on August 12, 2020).

The issue of data retrieval is rather more complex since this problem increases as the amount of data increases. The larger the library and the larger the library collection, the harder it becomes to lay one's hands on exactly the right item. Data can get lost or corrupted in an overloaded system. Additionally, as noted in the solution of networked storage devices, the added complexities of timely access and retrieval become more complex. The speed at which information can be retrieved and sent for processing can become an issue when data are widely dispersed on the network because of the basic laws of physics and distance. One way to address this issue is to use AI systems to manage where, how, and when to access the information. Information discovery has long been a land of algorithms for discovery and retrieval, so much so that it has become almost an expectation.

## MACHINE READING

Reading involves many subfields of computer science, but this is another area in which there has been significant progress in machine learning. Image processing, character recognition, and natural language processing are all elements of this subdomain, each contributing to the ability of machines to absorb content. Of course, when these processes have failed, humans have provided a helping hand by "reading" particularly difficult texts via technologies like text-based reCAPTCHA systems,[7] and there have consequently been significant advances in the accuracy of machine reading in the past 10 years.

Consider optical character recognition (OCR). In a work done by the National Library of Australia's Newspaper Digitisation Program from 2007 to 2008, an error rate of between 1 and 2 percent was considered "good OCR accuracy," but this also included manual correction. In 2007 the Koninklijke Bibliotheek (the Dutch National Library) conducted market research with a dozen OCR contractors, and the worst-performing vendor's service reported only 68 percent character accuracy.[8] Today error rates on typewritten pages are consistently so high that error rates are not even reported. A handwritten text, however, is a far more difficult OCR problem because of inconsistencies

7  Gugliotta, Guy. March 28, 2011. "Deciphering Old Texts, One Woozy, Curvy Word at a Time." *New York Times*. https://www.nytimes.com/2011/03/29/science/29recaptcha.html
8  Klijn, Edwin, January/February 2008. "The Current State of Art in Newspaper Digitisation: A Market Perspective." *D-Lib Magazine* 14 (1/2), doi:10.1045/january2008-klijn

in character formation, style, and flourishes in handwritten text, though systems are reported to be above 99 percent in some cases thanks in part to AI.[9]

These advances allowed for greater digitization of print collections to proceed at a rapid pace and at a lower cost. Following the digitization efforts of projects such as Project Gutenberg,[10] Google Books,[11] and the Internet Archive,[12] as well as the rapid expansion of open access content, there is a much larger corpus of text for training AI systems and to test their capacities. With these digitized corpuses of content, AI systems can iteratively improve their comprehension of human text, the rules for how characters, words, and grammar are used to communicate. This AI capacity will serve to allow for greater application of reading texts in a variety of contexts.

For example, the International STM Association recently reported that the growth of published journal content correlates closely with the increased number of researchers,[13] and it is unlikely that there will be a decrease in the growth of published research anytime soon. This has led to information overload and the feeling that many researchers are overwhelmed by keeping abreast of the pace of publication in their domain. Tools that can help researchers to consume this ever-increasing corpus of work would certainly be helpful. Machine reading can support researchers in this fashion, particularly, if it can be paired with an AI-based rendering of a summary. Some initial work on this, notably by Microsoft[14, 15] the Google Brain team, and the

---

9   Weiss, Rachel. June 27, 2019. "Have We Solved the Problem of Handwriting Recognition?" Website: https://towardsdatascience.com/https-medium-com-rachelwiles-have-we-solved-the-problem-of-handwriting-recognition-712e279f373b (accessed August 12, 2020).

10   https://www.gutenberg.org

11   https://www.google.com/googlebooks/about/

12   https://archive.org/scanning

13   Johnson, Rob, Watkinson, Anthony, and Mabe, Michael. October 2018. *The STM Report: An Overview of Scientific and Scholarly Publishing*, fifth edition. https://www.stm-assoc.org/2018_10_04_STM_Report_2018.pdf

14   Wiggers, Kyle. November 6, 2018. "Microsoft Develops Flexible AI System That Can Summarize the News. *Venture Beat*. https://venturebeat.com/2018/11/06/microsoft-researchers-develop-ai-system-that-can-generate-articles-summaries/

15   Fernandes, Patrick, Allamanis, Miltiadis, and Brockschmidt, Marc. February 3, 2021. *Structured Neural Summarization*. Preprint. https://arxiv.org/pdf/1811.01824.pdf

Imperial College of London AI reading and summarization service[16, 17] gives a glimpse of how these services might expand. Beyond this, machine reading can also support translation as part of these reading services. As the scholarly community becomes more international and more multilingual, machine reading services might not only be able to read but also to translate content to facilitate a greater international engagement in the sciences.

## AI UNDERSTANDING

When assessing whether a person understands what they are doing in a scientific context, we often ask them to "show their work." A person with a good memory could possibly repeat the first 100 of the Fibonacci sequence without understanding how to generalize the calculation as $x_n = x_{n-1} + x_{n-2}$ or how to determine the next item in the sequence beyond those memorized. The ability to recall 100 numbers in order certainly requires a certain sort of intelligence. Indeed, one of the challenges parents of a certain age have recently faced is how to help elementary school children with their mathematics, which is centered around process rather than memorization. Committing to memory, "12 times 12 equals 144" is one approach to producing the result but understanding how to multiply the two figures together to identify the result is potentially better.

Testing in schools has moved away from storage and retrieval (i.e., memorizing and recalling things) to something closer to describing your process and understanding how processes derive the answer. Since every child has access to a calculator, this process-based approach makes a lot of sense and allows for expressions of a higher level of intelligence. Similarly, computers' recall of information has far outpaced human capacity. Entire libraries worth of information can be stored on an external hard drive and retrieved in fractions of a second. More critical than retrieval, however, is understanding what is thereby retrieved and consumed.

16 Wiggers, Kyle. December 23, 2019. "Google Brain's AI Achieves State-of-the-Art Text Summarization Performance." https://venturebeat.com/2019/12/23/google-brains-ai-achieves-state-of-the-art-text-summarization-performance/

17 Zhang, Jingqing, Zhao, Yao, Saleh, Mohammad, and Liu, Peter J. 2020. "PEGASUS: Pre-training with Extracted Gap-Sentences for Abstractive Summarization." *Proceedings of the 37th International Conference on Machine Learning*, Vienna, Austria, PMLR 119, 2020. https://arxiv.org/pdf/1912.08777.pdf

An interesting example of this move from storage and retrieval to understanding is how to assess whether a piece of content is infringing copyright. At the moment, there are robust arguments around whether a piece of content can be shared in a digital environment. There are a variety of content-filtering technologies, such as ContentID[18] or automated content review[19] on social media platforms, which are AI-understanding-based. There have been some applications of AI or machine learning to identify copyrighted content that is uploaded to social media sharing sites. These systems are reasonably good at identifying content through processes such as digital fingerprinting[20] and pattern matching. To a certain extent this is akin to reading, but it is not quite understanding.

There are dozens of perspectives on what does or does not constitute the fair use of copyrighted content. To make a thorough determination of this question normally requires a case-by-case analysis that takes a variety of contextual awareness attributes into consideration. Copyright is not a simple binary system, and this makes an algorithmic solution to such a question extremely challenging and unlikely to yield satisfying results. Human experience is far too ambiguous and contextual to allow us to distill our answers to questions to a simplistic "yes" or "no," or "black" or "white." AI-based automated review processes, however, consistently apply a simplistic and determinative approach to profoundly complex issues. This approach can have its value in scaling a solution, whereby the "simpler" problems are addressed programmatically by the AI system, and a human-led approach is focused on those more complex issues.

As noted, machine learning is often based on a probabilistic approach to understanding, yet humans are challenged to operate in the same way. In part because of this, in our present environment we nonetheless place a great

---

18  "How ContentID Works." Website: https://support.google.com/youtube/answer/2797 370?hl=en (accessed September 12, 2020).

19  Kasperiuniene, J., Briediene, M., and Zydziunaite, V. 2020. "Automatic Content Analysis of Social Media Short Texts: Scoping Review of Methods and Tools." In Costa A., Reis L., and Moreira A. (eds.), *Computer Supported Qualitative Research. WCQR 2019. Advances in Intelligent Systems and Computing*, vol. 1068. Cham: Springer. https://doi.org/10.1007/978-3-030-31787-4_7

20  Milano, Dominic. *Content Control: Digital Watermarking and Fingerprinting*. White Paper. https://www.digimarc.com/docs/default-source/technology-resources/white-papers/rhozet_wp_fingerprinting_watermarking.pdf (accessed September 12, 2020).

deal of trust in the "black box" approach to problem-solving. Since we do not actually understand what is being done within the black box, we don't necessarily know how the machine arrived at its results. We may think, for example, that the AI is looking at the image of a cancer cell and matching the characteristics and patterns that make it resemble cancer; yet there may be another characteristic of the image that the computer finds is more correlated with the right result, such as whether the cell is dyed a particular way or the date the image was taken.

An example of this is work done by Janelle Shane examining how an AI trained on photos of sheep seemed only to be good at identifying white animals in a lush green landscape, not the actual sheep in the photo.[21] The characteristic of a white thing against a green background does not necessarily mean "sheep" or even an animal is represented. Perhaps machine learning is understanding something and making inferences based on the data that are very often correct, but this does not mean that they are comprehending what we want or expect them to comprehend. The more out of the ordinary the image, the more likely the computer is to fail in its recognition.

Humans spend a great deal of time online answering questions and helping to support the present world of machine understanding. The CAPTCHA system[22] had initially been designed to support machines that could not easily read warped, confusing, or misspelled text. Today, the CAPTCHA system is more often used to identify images, such as traffic lights, buses, or various traffic scenarios. Clearly, humans are still providing the training data for the increasing demands of AI in autonomous vehicles,[23] much as they had done in the early days of text reading and OCR processing. Over time this will improve, but we must continue to monitor the edge cases to know if we are simply being amazed by the power of recall and not by *actual* understanding.

---

21  Shane, Janelle. March 2, 2018. "Do Neural Nets Dream of Electric Sheep?" Website: https://aiweirdness.com/post/171451900302/do-neural-nets-dream-of-electric-sheep (accessed on August 10, 2020).

22  "The reCAPTCHA Project—Carnegie Mellon University CyLab." www.cylab.cmu.edu. Archived from the original on October 27, 2017.

23  Ganjoo, Shweta. January 21, 2019. "Do You Know You Are Training Google Self-Driving Cars So They Don't Kill People? Yes, by Solving Captcha." *India Today*. https://www.india today.in/technology/features/story/do-you-know-you-are-training-google-self-driving-cars-so-they-don-t-kill-people-1435604-2019-01-21 (accessed on September 1, 2020).

## AI SYNTHESIS

One element of the scholarly research process that shows the highest potential for AI and machine learning has to do with the synthesis of information. Machine learning and AI systems have potential in consuming vast stores of information, absorbing it, and then drawing inferences and connections about that information. We are starting to see significant results in this area, such as the identification of images of black holes,[24] scanning the eye to probe retinal disease,[25] and analyzing the MRI images of infants' brains to assess the risk of autism.[26] The absorption of vast amounts of information and the identification of unforeseen or even unexpected connections among disparate facts are the features of AI synthesis that are showing promise.

At the moment, AI synthesis is generally performing within specific domains since most AI systems are trained with a particular purpose in mind. A specific (or narrow) AI system can focus on neuroscience, or cancer detection, or astrophysics, but connecting information from the domains of physics, imaging, and cognition to develop a unifying theory is at present impossible. It is important to recognize that key scientific discoveries over the centuries have occurred at the intersection of existing domains such as Watson and Crick (Genetics, Physics, and Microbiology), Charles Darwin (Natural History, Biology, and Geology), or Jagadish Chandra Bose (Physics, Chemistry, and Botany). Again, much like the AI synthesis discussed earlier, the potential to support human advances is probably the key here as well. Supporting a human's understanding of these connections and helping to pursue the most potentially fruitful for more in-depth exploration will be useful.

Given large enough datasets, sufficient processing power, and sufficiently robust analytical approaches to data synthesis, new connections among domains is likely. Historically, these connections have happened when

---

24  Hardesty, Larry. June 6, 2016. "A Method to Image Black Holes." MIT News. Website: https://news.mit.edu/2016/method-image-black-holes-0606 (accessed on September 1, 2020).

25  Ting, D.S.W., Pasquale, L.R., Peng, L., et al. 2019. "Artificial Intelligence and Deep Learning in Ophthalmology." *British Journal of Ophthalmology* 103: 167–175. http://dx.doi.org/10.1136/bjophthalmol-2018-313173

26  Hazlett, H., Gu, H., Munsell, B., et al. 2017. "Early Brain Development in Infants at High Risk for Autism Spectrum Disorder." *Nature* 542: 348–351. https://doi.org/10.1038/nature21369

polymaths with interests in multiple domains sought to reconcile or transfer their experiences from one domain to another. It is realistic to assume that machines, with their significantly larger storage, recall, and statistical analysis capabilities, will be able to identify the key points of interaction across domains. However, identifying these potentially synthetic areas also requires an element of creativity and understanding. It will be interesting to see whether AI systems or humans will be better at spotting those similarities and relationships. Perhaps the key will not simply be in "seeing" those connections but in recognizing their importance and noting them in the pursuit of other discoveries.

## AI CREATIVITY

At present there are a variety of applications of AI that purport to be creative, including the uses of GANs neural networks to produce artistic images,[27] musical works,[28] and even fictional texts.[29] Technically speaking, in each of these three examples, the GANs or AI "created" the work of art. However, three important elements of the creative process are still significantly controlled by humans. The first is the input, the second is the intention, and the third is the editorial selection of the outputs. Before employing an algorithm on a creative project, that algorithm must review a set of data that resembles the desired output to understand what the desired output might be. One could not train an AI to produce an English sonnet, if all it was provided was Korean business reports as training data. That gathering of data is a curatorial one, which relies on human understanding of what a sonnet is like and what is the corpus of things that are the reference for subsequent development. Once gathered, an AI will only tend toward the results that are governed by its algorithm and the rule set that has been established in that process, regardless

27 Jones, Kenny. June 18, 2017. "GANGogh: Creating Art with GANs." Website: https://towardsdatascience.com/gangogh-creating-art-with-gans-8d087d8f74a1 (accessed October 18, 2020).
28 Misasi, Charles Robert, Zehden, David, Wei, Thomas et al. May 22, 2019. "Generating Music with a Generative Adversarial Network." Website: https://medium.com/ee-460j-final-project/generating-music-with-a-generative-adversarial-network-8d3f68a33096 (accessed on October 18, 2020).
29 Martin, Eric. June 18, 2018. "Will DeepMind Use GANs to Write the Next Harry Potter?" Website: https://medium.com/predict/will-googles-deepmind-use-gans-to-write-the-world-s-next-harry-potter-bf6c3d283410 (accessed on October 18, 2020).

of how "free" it might seem. There are a variety of approaches to training a machine learning system, such as supervised learning, reinforcement learning, deductive learning, and so on, each with its own rules and approach. While there have been glimmers of creative solutions to problems—even "elegant" solutions in some cases[30]—all these AI solutions happen within constrained environments. The intention of solving a problem and then the approach that is used to discern a solution is almost always provided externally to the AI system by the creative agent that is managing the solution.

The final element of the creative process is editorial. How does an artist select among his or her completed works what to keep and what to discard? A finely crafted creative work is usually wrought and worked over many times before the artist considers it complete. While there are certainly exceptions, even the work an author considers complete is improved through editing and review. In a studio, musicians will record and remix a song many times to achieve what they regard as the finished work. But even after the authors and editors have done their work, the act of selecting the final objects for release and distribution is yet another level of curation. The final work is often just one of many photographs that were taken, musical tracks recorded, or papers written. An AI process that produced some sort of art is not likely to be evaluating the outputs and determining what should go onto the final exhibit. This final evaluation is done by human curators.[31]

Likewise, creativity involves more than simply rolling a contraption down a hill. In the 19th Century, when the U.S. Court systems were considering the application of copyright to photography, the distinction was drawn between the creative acts of selection and of manipulating the tool to create a work of art. A creative photographic work involves much more than simply opening and closing the shutter on a camera. AI systems are a tool, but creativity remains deeply embedded in the mind of the human curator.

---

30  Menick, John, 2016. "Move 37: Artificial Intelligence, Randomness, and Creativity." *Mousse Magazine* 55 + 53. Accessed via: http://johnmenick.com/writing/move-37-alpha-go-deep-mind.html (accessed on October 18, 2020).
31  Ganvas Studio run by Danielle Baskin is an example of this. See https://ganvas.studio

## CHALLENGES EVEN AN AI MIGHT NOT BE ABLE TO SOLVE

For all their possibilities and the opportunities AI systems might present, there are many reasons to be cautious about our notion of an AI-driven future. The first issue has to do with bias. As each system is trained on a set of data, the selection of those data is crucially important. The preponderance of certain data within the training dataset could be magnified by the machine processing of AI systems—for example, if one gathers data in Korea and in Miami, there will be a preponderance of certain data and an absence of others. Without understanding how a machine is processing these data, it is hard to detect bias simply by examining the result set. Beyond this, the AI system could reinforce unforeseen biases in the data in the result set.

Additionally, algorithmic thinking cannot discern the correct answer in many situations. As noted earlier, ambiguity is part of every human's life and intuition and experience may yield the best answer to a dilemma. Furthermore, there are many occasions both in machine learning and in human experience in which we determined the correct answer to a problem, but we arrived at it or by using a questionable, even erroneous, methodology. We might be satisfied with the initial result, thinking that the AI has solved the problem, when in fact it is "cheating" its way to the correct answer. While the AI may derive the appropriate answer the first few times, another set of data may leave us seriously disappointed as to how the first answer was derived. The more serious the consequences of the problem being solved using an AI, the more we want to be certain that the answers are reliable, which requires an understanding of how the answers were determined. Addressing this problem is in some ways predicated on understanding what the process is that is taking place within the "black box" of the algorithm. There has been some interesting exploration of auditable AI processes,[32] which should see greater exploration as AI applications advance.

---

32 Adadi, A., and Berrada, M. 2018. "Peeking Inside the Black-Box: A Survey on Explainable Artificial Intelligence (XAI)," *IEEE Access* 6: 52138–52160, doi: 10.1109/ACCESS.2018.2870052.

As much as we might hope it to be the case, machines will not be creative for us in the traditional ways we think of creativity. However, AI systems will help us to creatively test our theories, test our approaches at scale, and filter information in the world to discern the possible and impossible, the functional and the inoperable, and the potential and the impractical solutions. Machine learning is already doing this in some domains, and this will only continue.

## MACHINES WILL HELP US BE MORE PERSISTENT FASTER

Thomas Edison is said to have tried thousands of times to create a better filament to improve the function of the light bulb.[33] He was one among dozens—if not hundreds—of scientists who were testing various ways to improve the use of electricity as a functional light source. We may remember Edison as a creative and inventive personality, but another way to understand him is as a man driven obsessively to pursue a project until he found the right outcome. Persistence, therefore, may be the trait for which Edison may more rightly be remembered. Perhaps AI will support our creativity and allow us to test our theories more quickly? Machines will be able to absorb tremendous amounts of data or run millions of simulations to discern those paths that will definitely not work, those that may work, and those that are most likely to work. In effect, AI systems would allow us to be more creative—or at least more persistent—faster.

Increasingly, AI will become a ubiquitous tool for information scientists and scholars of all types. Applying the tools of AI to problem-solving will become second nature, as when we use the basics of computer technology, or how accustomed we have become to barking orders at our voice assistants. AI will cease to become a magical feature on the far scientific horizon and will enter the mundane day-to-day experience of most scholars. AI or various machine learning approaches will support scholars to determine the potential directions of research, analyze the resulting data, and assist the drafting of papers. After that, AI and machine learning systems can help with the vetting and submission processes that lead to publication. It will not seem like

---

33 "Edison's Light Bulb." The Franklin Institute. https://www.fi.edu/history-resources/edisons-lightbulb

magic, nor will it appear that any machines are contributing meaningfully to the processes of research and publication. Certainly, it will not be anything like having the cyborg "Data" from *Star Trek: The Next Generation* supporting one's research efforts! But no matter how far technology advances, we never quite end up living in the worlds that science fiction had envisioned. Regardless of how advanced AI might become in understanding our world, it will not be able to foresee the future, and that is a quality it will share with its human creators.

## REFERENCES

Adadi, A., and Berrada, M. 2018. "Peeking Inside the Black-Box: A Survey on Explainable Artificial Intelligence (XAI)," IEEE Access 6: 52138–52160, doi: 10.1109/ACCESS. 2018.2870052.

"Edison's Light Bulb." The Franklin Institute. https://www.fi.edu/history-resources/edisons-lightbulb

Fernandes, Patrick, Allamanis, Miltiadis, and Brockschmidt, Marc. February 3, 2021. Structured Neural Summarization. Preprint. https://arxiv.org/pdf/1811.01824.pdf

Frické, Martin, 2019. "The Knowledge Pyramid: The DIKW Hierarchy." Knowledge Organization 49 (1): 33–46. Also available in ISKO Encyclopedia of Knowledge Organization, eds. Birger Hjørland and Claudio Gnoli, http://www.isko.org/cyclo/dikw

Ganjoo, Shweta. January 21, 2019. "Do You Know You Are Training Google Self-Driving Cars So They Don't Kill People? Yes, by Solving Captcha." India Today. https://www. india today.in/technology/features/story/do-you-know-you-are-training-google-self-drivingcars-so-they-don-t-kill-people-1435604-2019-01-21 (accessed on September 1, 2020).

Gent, Edd. April 1, 2019. "Microsoft Is Building an All-in-One DNA Data Storage Device." Website: https://singularityhub.com/2019/04/01/microsoft-is-building-an-all-inone-dna-data-storage-device/ (accessed on August 12, 2020).

Gugliotta, Guy. March 28, 2011. "Deciphering Old Texts, One Woozy, Curvy Word at a Time." New York Times. https://www.nytimes.com/2011/03/29/science/29recaptcha.html

Hardesty, Larry. June 6, 2016. "A Method to Image Black Holes." MIT News. Website: https://news.mit.edu/2016/method-image-black-holes-0606 (accessed on September 1, 2020).

Hazlett, H., Gu, H., Munsell, B., et al. 2017. "Early Brain Development in Infants at High Risk for Autism Spectrum Disorder." Nature 542: 348–351. https://doi.org/10.1038/nature21369

"How ContentID Works." Website: https://support.google.com/youtube/answer/2797 370?hl=en (accessed September 12, 2020).

Johnson, Rob, Watkinson, Anthony, and Mabe, Michael. October 2018. The STM Report: An Overview of Scientific and Scholarly Publishing, fifth edition. https://www.stm-assoc. org/2018_10_04_STM_Report_2018.pdf

Jones, Kenny. June 18, 2017. "GANGogh: Creating Art with GANs." Website: https://towardsdatascience.com/gangogh-creating-art-with-gans-8d087d8f74a1 (accessed October 18, 2020).

Kasperiuniene, J., Briediene, M., and Zydziunaite, V. 2020. "Automatic Content Analysis of Social Media Short Texts: Scoping Review of Methods and Tools." In Costa A., Reis L., and Moreira A. (eds.), Computer Supported Qualitative Research. WCQR 2019. Advances in Intelligent Systems and Computing, vol. 1068. Cham: Springer. https://doi. org/10.1007/978-3-030-31787-4_7

Klijn, Edwin, January/February 2008. "The Current State of Art in Newspaper Digitisation: A Market Perspective." D-Lib Magazine 14 (1/2), doi:10.1045/january2008-klijn

Kryder's Law. https://en.wikipedia.org/wiki/Mark_Kryder#Kryder's_law_projection

Martin, Eric. June 18, 2018. "Will DeepMind Use GANs to Write the Next Harry Potter?" Website: https://medium.com/predict/will-googles-deepmind-use-gans-to-write-theworld-s-next-harry-potter-bf6c3d283410 (accessed on October 18, 2020).

Martin, George R. R. May 14, 2014. "Why I Still Use DOS." Website: https://www.bbc.com/news/technology-27407502 (accessed September 20, 2020).

Menick, John, 2016. "Move 37: Artificial Intelligence, Randomness, and Creativity." Mousse Magazine 55 + 53. Accessed via: http://johnmenick.com/writing/move-37-alphago-deep-mind.html (accessed on October 18, 2020).

Milano, Dominic. Content Control: Digital Watermarking and Fingerprinting. White Paper. [n.d] https://www.digimarc.com/docs/default-source/technology-resources/white-papers/ rhozet_wp_fingerprinting_watermarking.pdf (accessed September 12, 2020).

Misasi, Charles Robert, Zehden, David, Wei, Thomas et al. May 22, 2019. "Generating Music with a Generative Adversarial Network." Website: https://medium.com/ee-460j-final-project/generating-music-with-a-generative-adversarial-network-8d3f68a33096 (accessed on October 18, 2020).

Moore's Law. Wikipedia. https://en.wikipedia.org/wiki/Moore%27s_law

Rosenthal, David. December 13, 2016. "The Medium-Term Prospects for Long-Term Storage Systems." https://blog.dshr.org/2016/12/the-medium-term-prospects-for-long-term.html

Shane, Janelle. March 2, 2018. "Do Neural Nets Dream of Electric Sheep?" Website: https://aiweirdness.com/post/171451900302/do-neural-nets-dream-of-electric-sheep (accessed on August 10, 2020).

"The reCAPTCHA Project—Carnegie Mellon University CyLab." www.cylab.cmu.edu. Archived from the original on October 27, 2017.

Ting, D. S. W., Pasquale, L. R., Peng, L., et al. 2019. "Artificial Intelligence and Deep Learning in Ophthalmology." British Journal of Ophthalmology 103: 167–175. http://dx.doi.org/10.1136/bjophthalmol-2018-313173

Weiss, Rachel. June 27, 2019. "Have We Solved the Problem of Handwriting Recognition?" Website: https://towardsdatascience.com/https-medium-com-rachelwiles-havewe-solved-the-problem-of-handwriting-recognition-712e279f373b (accessed August 12, 2020). 10 https://www.gutenberg.org 11 https://www.google.com/googlebooks/about/ 12 https://archive.org/scanning

Wiggers, Kyle. November 6, 2018. "Microsoft Develops Flexible AI System That Can Summarize the News. Venture Beat. https://venturebeat.com/2018/11/06/microsoft-researchers-develop-ai-system-that-can-generate-articles-summaries/

Wiggers, Kyle. December 23, 2019. "Google Brain's AI Achieves State-of-the-Art Text
    Summarization Performance." https://venturebeat.com/2019/12/23/google-brains-
    ai-achieves- state-of-the-art-text-summarization-performance/
Zhang, Jingqing, Zhao, Yao, Saleh, Mohammad, and Liu, Peter J. 2020. "PEGASUS: Pre-
    training with Extracted Gap-Sentences for Abstractive Summarization." Proceedings of
    the 37th International Conference on Machine Learning, Vienna, Austria, PMLR 119,
    2020. https://arxiv.org/pdf/1912.08777.pdf

# ABOUT THE AUTHORS

Amy Brand is director and publisher of the MIT Press. Active in the MIT Media Lab and the Knowledge Futures Group.

Todd A. Carpenter is Executive Director of the National Information Standards Organization (NISO).

Catherine Nicole Coleman is Digital Research Architect, Stanford University Libraries.

Haris Dindo is Chief Technology Officer at SHS Asset Management. He was formerly Chief Data Scientist at Yewno, Inc.

Ruggero Gramatica is Founder and CEO of Yewno, Inc.

Daniel W. Hook is CEO of Digital Science, Visiting Professor at Washington University in St Louis, Co-chair at the Research on Research Institute, and Policy Fellow at the Centre for Science and Policy at Cambridge University.

Michael A. Keller is Vice Provost and University Librarian & Vice Provost for Teaching and Learning at Stanford University Libraries.

Ruth Pickering is Co Founder and Chief Strategy and Business Development Officer of Yewno, Inc.

Simon J. Porter is Director of Innovation at Digital Science.

James W. Weis is Founder and CEO of Nest.Bio. He has been active in the MIT Media Lab, Knowledge Futures Group, and MIT Computational & Systems Biology.